BIOCHEMISTRY RESEARCH TRENDS

GENETIC, HEALTH AND ENVIRONMENTAL IMPACTS CAUSED BY GLYPHOSATE IN ECUADOR

BIOCHEMISTRY RESEARCH TRENDS

Additional books in this series can be found on Nova's website under the Series tab.

Additional e-books in this series can be found on Nova's website under the e-book tab.

BIOCHEMISTRY RESEARCH TRENDS

GENETIC, HEALTH AND ENVIRONMENTAL IMPACTS CAUSED BY GLYPHOSATE IN ECUADOR

CÉSAR PAZ-Y-MIÑO
AND
ANDRÉS LÓPEZ-CORTÉS

nova publishers

New York

For permission to use material from this book please contact us:
Telephone 631-231-7269; Fax 631-231-8175
Web Site: http://www.novapublishers.com

LIBRARY OF CONGRESS CATALOGING-IN-PUBLICATION DATA

ISBN: 978-1-63321-618-1
Library of Congress Control Number: 2014945908

Published by Nova Science Publishers, Inc. † New York

Dedicated to

Ariane (CPyM)
Pato, Paty, Richard, Nathy and Caro (ALC)

CONTENTS

About the Authors

César Paz-y-Miño
Dean of Biomedical Research Institute
Faculty of Health Science
Universidad de las Americas
Quito, Ecuador
E-mail: cesar.pazymino@udla.edu.ec
Twitter: @CesarPazyMino

Andrés López-Cortés
Biomedical Research Institute
Faculty of Health Science
Universidad de las Americas
Quito, Ecuador
E-mail: andres.lopez@udla.edu.ec
Twitter: @Andres_Lopez_C

ACKNOWLEDGMENTS

PHOTOGRAPHER:

Lucas Bustamante-Enríquez
The photographs of this book are about the biodiversity of the Ecuadorian Amazon region.

PEER REVIEW:

Francisco Benítez-Capistros
Laboratory of Plant Biology and Nature Management,
Vrije Universiteit Brussel.
Laboratory of Systems Ecology and Resource Management,
Université Libre de Bruxelles

Juan Carlos Ruiz
Laboratory of Molecular Biology,
National Cancer Institute,
Oncological Hospital Solon Espinoza Ayala

PREFACE

Glyphosate is one of the most widely used chemical products worldwide and its manufacturer considers it the safest herbicide for human health around the world under the stated technical conditions: a direct fumigation on weed, concentration use of between 1 to 3%, and low dose and low exposure time on the part of the person who is handling the herbicide.

In Ecuador the aerial spraying that the Government of Colombia carried out from 2000 to 2007, in connection with the Colombian plan to combat coca, opium poppy, and marijuana crops, affected the communities located along the Colombian-Ecuadorian international border. This caused both countries to take up divergent diplomatic, scientific-technical, ethical, and governmental positions which ended up with a State problem.

Both countries, by means of their governments, appointed scientific commissions which did not reach an agreement during bilateral discussions. For Ecuador, the problem lies in the fact that the use of glyphosate is anti-technical: the aerial spraying was carried out from a great height in a repetitive manner which used a combination of herbicides that, besides containing glyphosate as an active agent, contained Cosmo-Flux as coadjuvant, and polyoxyethylene amine (POEA) as surfactant. Therefore, the chronic exposure to the environment (water, plants, and land), animals, and people resulted in important changes and problems due to the indirect effects that affected the Ecuadorian side of the border. The effects were observed and determined, according to the testimony of the people from the affected areas who commented on the deaths of their animals, plants, and water, as well as the denunciations made at the Ombudsman's Office. The effects were verified by several national and international commissions, including the United Nations. At the same time, Ecuadorian doctors reported on changes in the health patterns of the people exposed to the aerial spraying with the herbicide compound. Gastrointestinal, skin, respiratory, eye, and psychological problems and disorders increased, according to the Ecuadorian Scientific Commission (Comisión Científica Ecuatoriana), appointed by the President of Ecuador, Rafael Correa, and his minister of Foreign Affairs at that time, Maria Fernanda Espinosa.

The research conducted by the Ecuadorian Scientific Commission was captured in the book "The Plan Colombia's Aerial Spraying System and its Impact on the Ecosystem and Health in the Ecuadorian Border" (2007) which includes the different evidence and damage that Ecuador claims to have experienced. As result of the documentation collected by Ecuador, the country decided to file an international lawsuit before The Hague after alleging that the fumigation caused serious damage to the Ecuadorian health and environment.

There is currently plenty of scientific information that questions the innocuousness of glyphosate by itself and combined with other products, especially an herbicide compound with a high concentration, and an application method for which there is no comparison on the planet whatsoever. The analyses of the damage caused by glyphosate to the genetic material prove the DNA degradation, by means of the comet assay test and chromosome breakage, turning it into a genotoxic agent. Genotoxics are involved in at least three important biological phenomena: cancer development, miscarrying, and infertility in the exposed individuals.

The aerial spraying with the herbicide compound containing glyphosate stopped in January 2007 and the experience the Ecuadorian people gained from this weird and new factor that altered the biology along the border has enriched academic knowledge, as well as the wisdom in the communities located in the affected areas.

Glyphosate has a short lifespan; its manufacturer talks about three months, but the analyses on the environmental destination of this herbicide are able to detect it up to two years later, especially in soils rich in iron. Whatever the case may be, it is obvious that the aerial spraying caused problems along the border, which have been studied by several organizations over time. After becoming aware of this situation, the Biomedical Research Institute at the Universidad de las Américas outlined a project that was approved by the National Science and Technology Secretariat (Secretaría Nacional de Ciencia y Tecnología-SENACYT), currently known as the National Higher Education, Science, Technology, and Innovation Secretariat (Secretaría Nacional de Educación Superior, Ciencia, Tecnología e Innovación-SENESCYT), in order to evaluate the impact of the aerial spraying with glyphosate on the Northern border, and to analyze for two years an array of genetic, physical, and psychological factors that define the results that this book gathers.

By means of this new book, which includes many actors and researchers, we intend to provide scientific information related to the pollution, precaution, protection, and remediation mechanisms, as well as the potential effects caused by the indirect and direct aerial spraying on the Ecuadorian communities settled along the border with Colombia. For a better understanding of the impact caused by the herbicide cocktail, the SENACYT created the Ecotoxicology Network made up of the Universidad de las Américas, the Ecuadorian Natural Sciences Museum, and the Escuela Politécnica Nacional, which focuses on exchanging information that these four institutions will obtain regarding the same topic: aerial spraying with the herbicide cocktail and its effects on the Ecuadorian territory.

The results of the project, together with a video, were previously presented at the 13th Pacific Basin Consortium Meeting, called "Environmental Exposures in the Era of Climate Change", which took place in November 2009 at the Telethon Institute for Child Health Research in Perth, Australia, titled "Social, Health and Genetic Diagnoses of the Zone of Influence Affected by Glyphosate Aerial Spraying in the North Ecuadorian Border"; the formal scientific work is available on the magazine Reviews on Environmental Health, Vol. 26, No. 1, January 2011, titled "Baseline Determination in Social, Health, and Genetic Areas in Communities Affected by Glyphosate Aerial Spraying in the Northeastern Ecuadorian Border".

Once the project under my supervision is finished and the results that we present today are extended and analyzed in this book, it may be considered a "baseline" study of the genetic, cytogenetic, social, health, and environmental conditions from the communities in the province of Sucumbios bordering Colombia.

This book provides information on the effects of this genotoxic on the DNA structure and chromosomes; the mechanism of action of glyphosate and its increasing action by means of surfactants, coadjuvants, and other pollutants; the international toxicological studies of pesticides in humans and other organisms; the Ecuadorian Amazon biodiversity threatened by the aerial spraying, as well as the effects on crops, animals, and water; the origin and life conditions of the communities; the study of chromosome alterations and DNA genetic polymorphisms related to large or little susceptibility to glyphosate; the valuation of the nutritional status; the mother-child health and the impact of the aerial spraying on the psychological condition; the environmental diagnosis by means of the analysis of glyphosate and AMPA in soil samples;

and the parasitological and microbiological analysis of water samples. To finalize, this book is also a good source of bibliographic information on antimutagenic agents and physical-chemical, enzymatic, and oxidation mechanisms used in other countries in order to mitigate and remediate the presence of glyphosate in the environment; this can be helpful in Ecuador if the indiscriminate use of this pesticide continues.

The importance of conducting a strictly scientific research of wide range in order to determine a baseline for the genetic, health, and environmental areas, as well as the importance of continuing with short and long-term prospective studies will allow both public and private institutions to evaluate in a more realistic manner the situation of the Ecuadorian communities, and to carry out social, psychological, nutritional, and environmental projects, with the aim of improving the social-economic, nutritional, health, psychological, and environmental levels of the Ecuadorian communities bordering Colombia.

César Paz-y-Miño
Quito, Ecuador
2015

Chapter 1

HERBICIDE PACKAGE-MIX

ABSTRACT

This chapter discusses glyphosate, its chemical structure, and its mechanism of action, by means of the inhibition of the enzyme 5-enolpyruvylshikimate 3-phosphate synthase. This chapter also focuses on the role played by the surfactant, coadjuvant, and pollutant components that reinforce the mechanism of action of the organophosphorus pesticides, such as: Roundup, polyacetylene amine, Cosmo-Flux 411F, and N-Nitroso Glyphosate. On the other hand, this chapter deals with plant genetic resistance to glyphosate, transgenic plants tolerant to glyphosate, and its serious ecological and environmental consequences due to the adaptive capacity that the plants have because of the farmers' daily use of this component and its use in anti-narcotic programs.

1. GLYPHOSATE

The N-(phosphonomethyl) glycine (C3H8NO5P) was reported as an herbicide in 1971. This substance, commonly known as glyphosate, is an organophosphorus, systemic, non-selective, broad-spectrum, and effective herbicide to control and kill any type of annual, biennial, and perennial plants, including woody plants. Additionally, glyphosate is the active component of several commercial pesticide formulations and herbicides that are manufactured and distributed worldwide (Smith & Oehme, 1992; Duke & Powles, 2008).

At room temperature, glyphosate is a clear white liquid, made up of positive and negative charges. In water, the hydrogen atoms of the carboxylic acid (COOH) and the phosphate group (PO_2H_2) have the capacity to associate (-COOH) or dissociate ($-COO^- + H^+$), depending on the PH level of the solution that varies from 5 to 9 (USDA, 2003). Technically, glyphosate has a micropollutant called N- nitrosoglyphosate (NNG). The physical, chemical, and biochemical properties of glyphosate are detailed in Table 1.1.

There are currently 35 glyphosate commercial formulations in the market that are registered to be used as an herbicide in farming. All these formulations commonly used glyphosate as salt; isopropylamine salt of N-(phosphonomethyl) glycine is the most used one. The solubility of glyphosate is relatively high in water and the ionic nature of this herbicide slows down the infiltration in the plants with hydrophobic cuticles. Consequently, glyphosate is commonly formulated with surfactants that decrease the surface tension of the solution and therefore, the infiltration capacity in tissues increases (WHO, 1994; Giesy et al., 2000).

Glyphosate is one of the most used herbicides in farming worldwide. It is part of dozens of products prepared by several pesticide manufacturers. The sale of products with glyphosate has generated a profit of approximately $1.2 billion dollars a year, and encompasses 60% of the sale related to non-selective herbicides (Zeneca, 1995). This herbicide is mostly used in the United States with around 51,000 tons, and Europe with 20,000 tons (Acquavella et al., 2004; Kiely et al., 2004).

1.1. Mechanism of Action

The mechanism of action of glyphosate is mainly based on the inhibition of the enzyme 5-enolpyruvylshikimate 3-phosphate synthase (EPSPS), generating delay or collapse of the metabolic route of shikimato, which is involved in the synthesis of essential aromatic aminoacids, such as phenylalanine, tyrosine, tryptophan, and this way, the growth of plants and microorganisms is stopped (USDA, 2003; Mladinic et al., 2009). Starting from the shikimic acid, other aromatic products are produced, such as lignin, alkaloids, flavonoids, benzoic acids, and phytohormones that are typical for the secondary metabolism such as allelochemics. In fact, some 20% of the established carbon is used in this metabolic route (CCE, 2007).

Table 1.1. Glyphosate physicochemical and biochemical properties

CAS #:	**1071-83-6**
Chemical nomenclature:	N-(phosphonomethyl) glycine
Chemical classification/use:	Phosphanoglycine herbicide
Water Solubility:	12 g/L (77°F)
	900 g/L, amine salt (Knisel et al., 1992)
	11.6 g/L (Schuette, 1998)
Solubility in different solvents:	Insoluble in organic compounds
Molecular weight:	169.07 (Budavari, 1989)
Melting point:	365° F (decomposes at 390.2° F)
Vapor pressure (mm Hg):	< 7 x 10-9 mm Hg (77°F)
Value of dissociation:	2.27 (pK_a)
Log Kow:	-3.39 (pH 1.77)
	-4.38 (pH 4.61)
	-4.85 (pH 6.86)
	-4.14 (pH 9.00)
	-3.5 (Schuette, 1998)
Water-Land Partition coefficient K_d:	61 g/m^3 (Schuette, 1998)
	2100 (500-2600)
Land Absorption Koc (L/kg):	554-34000 (Piccolo et al., 1994)
	2.600-4.900 (Glass, 1987)
	8 a >500000 (Gerritse et al., 1996)
	54 (Knisel et al., 1992)
CAS #:	**1071-83-6**
Lifespan on leaves (days):	1.6 (Thompson et al., 1994)
	8-10 (Feng & Thompson, 1990)
	10.6-26.6 (Newton et al., 1984)
	2.5 (Knisel et al., 1992)
	46 (Siltanen et al., 1981)
Lifespan on land (days):	20-40
	< 60
	45-60 (Feng & Thompson, 1990)
	29-40 (Newton et al., 1984)
	30-40 (Smith & Aubin, 1993)
	47 (Knisel et al., 1992)
	22.1 aerobic; 96.4 anaerobic (Schuette,1998)
Lifespan in water (days):	50-70 (US EPA/ODW, 1992) 42-70 (Reinert & Rodgers, 1987) > 35 (Schuette, 1998)

Due to the fact that the metabolic route of the shikimic acid does not exist in animals, the acute glyphosate toxicity is low. This herbicide may interfere with biochemical mechanisms like the uncoupling of cellular processes of oxidative phosphorylation and the inhibition of the hepatic mixed function oxidase system. Oxidative phosphorylation is an essential metabolic process whereby the metabolic energy that comes from nutrient oxidation is transferred and stored in high-energy phosphate bonds; the uncoupling of this process results in energy loss in organisms and consequently, their death. The symptoms of dysfunction in the processes of oxidative phosphorylation are increased heart rhythm, breathing problems, fever, metabolic acidosis, and weight loss (ATSDR, 2001). The inhibition of the hepatic mixed function oxidase system acts in the metabolism of foreign and toxic compounds. The symptoms of intoxication depend on the doses and the exposure time. Nevertheless, the products containing glyphosate also have other toxic compounds.

The main metabolite for glyphosate degradation in land is the aminomethyphosphonic acid (AMPA), which is also toxic. It starts as an AMPA, then it goes to methylamine, and ends as a formaldehyde, a well-known carcinogen (Cox, 1991 y 1995; Dinham, 1999).

2. Roundup: Surfactant, Coadjuvant, and Pollutants

The World Health Organization recommends glyphosate to be classified in accordance with its danger as Class IV: “Very likely to be dangerous for normal use”. The U.S. Environmental Protection Agency (US EPA, 1991) considers glyphosate as Class III: “mild irritant compound” (WHO, 2004). The characteristics to be considered are detailed in Table 1.2.

Every pesticide has, in addition to its active component, other substances focused on facilitating its use or increasing its efficiency. In general these components (called “inert”) are not detailed on the labels of the products. In the case of herbicides like glyphosate, many inert components have been identified. In order to help glyphosate infiltrate in the plants’ tissue, most of the commercial formulations contain surfactants. Therefore, the toxicological characteristics of the commercial products are different from those found in glyphosate only.

Table 1.2. List of inert products based on their harmful properties

	Harmful properties
List 1	Inert products that have shown cancer potential, effects on cancer development, and neurotoxic effects, are a risk to the ecology, and are urgent in terms of statutory actions.
List 2	Highly urgent products due to the fact that the toxicity data are subjective but not conclusive; the chronic effects on health are possible, or have a similar structure to the products in List 1.
List 3	Non-urgent products in terms of statutory actions since the toxicity data or the evaluation of their chemical structure are not a concern regarding toxicity or risk when using the products.
List 4	Inert components generally accepted as safe products.

Pesticide manufacturers use a long list of inert components that are found in different glyphosate-based commercial formulations. Every one of these inert compounds generates symptoms of acute toxicity in human beings when the dose is high and the exposure time is considerable. The symptoms caused by the inert components are detailed in the following table:

Table 1.3. Inert components and their symptoms

	Symptoms
Ammonium sulfate	Eye irritation, nausea, diarrhea, respiratory allergies, and irreversible eye damage after excessive exposure.
Benzisothiazolinone	Eczema, skin irritation, allergic photoreactions in sensitive individuals.
3-iodine-2-propynyl butylcarbamate	Serious eye irritation, frequency of abortion, skin allergies.
Isobutane	Nausea, nervous system depression, dyspnea.
Methylpyrrolidone	Serious eye irritation, abortion, and low birth weight in animals.
Pelargonic acid	Eye, skin, respiratory tract irritation.
Potassium hydroxide	Irreversible eye damage, deep skin ulceration, serious digestive tract ulceration, lung irritation.
Sodium sulfite	Eye and skin irritation with nausea and diarrhea, skin allergy, and serious allergic reactions.

There are several liquid formulations containing glyphosate, such as Aquamaster, Glyfos Aquatic, Roundup Custom, Rodeo. Other commercial formulations containing glyphosate and surfactants are Credit, Glyfos, Glyfos Pro, Glyfos X-TRA, Honcho, Mirage, Razor, Roundup, Roundup PRO. The most used herbicide formulation is called Roundup, a chemical formulation of which is registered by the company Monsanto.

Roundup is a water solution made up of glyphosate as its active component, polyoxyethylene amine (POEA) as surfactant, Cosmo-Flux 411F as coadjuvant, and N-Nitroso glyphosate (NNG) as micropollutant (Doliner, 1991; Tsui & Chu, 2003).

2.1. Polyoxyethylene Amine (POEA)

According to the information provided by Martin Jelsma, expert in international policy on drugs and researcher associated with the *Transnational Institute* (TNI), the company Monsanto has confirmed that the active component of Roundup Ultra is the same glyphosate of the traditional commercial Roundup. According to *Farm Chemicals Handbook 2000*, glyphosate is chemically identified as an Isopropylamine salt of N-(phosphonomethyl) glycine. The Roundup Ultra formulation has the following composition.

The polyethoxylated part means that it has multiple units (CH_2CH_2O), where C, H, and O are carbon, hydrogen, and oxygen. A "tallowamine" comes from the animal fat (tallow) and is from an amine functional group. Amines are weak bases and their ammonium salts are weak acids, but they can cause corrosion in skin due to their acid-based properties (Nivia, 2001).

The POEA is an anionic tensioactive agent used in the Roundup commercial formulation in an approximate proportion of 15% (150g/L). It comes from animal fat and is chemically known as a polyoxyethylene tallow amine. The anionic surfactants are often used in anionic detergents and soaps (Onil, 2001).

This surfactant contains an array of tallow acids, including the following: oleic (37-43%), palmitic (24-32%), stearic (20-25%), myristic (3-6%), and linoleic (2-3%), as well as small quantities of cholesterol (Budaveri, 1989).

Table 1.4. Chemical Formulation of Roundup Ultra

Roundup Ultra	
41%	Isopropylamine salt of glyphosate (active component)
44,5%	Water
14,5%	A phosphate ester neutralized polyethoxylated tallowamine mixture; Surfactant that increases the power of infiltration of the active component

There are several formulations of glyphosate that have POEA at levels greater than 20% (200 g/L). The POEA is a surfactant that is used for glyphosate to go through plant cuticles, since it has hydrophobic characteristics, which complicates the absorption of the chemical substance. The so-called "inert components" perform as solvents and moisturizers that can increase permeability and go through the plant cuticle that provides a barrier (CCE, 2007).

Some research on glyphosate shows that adding surfactants may increase the effects of herbicides and produce phytotoxicity (O'Sullivan et al., 1981; Sundaram, 1990; Laerke, 1995; Riechers et al., 1995; De Ruiter & Meinen, 1996; Sundaram et al., 1996; Denis & Delrot, 1997; Miller et al., 1998). Additionally, several studies conclude that the effects of surfactants may become more harmful for plants and animals than an herbicide itself (Chang et al., 1999; Garry et al., 1999; Lin & Garry, 2000).

The POEA has greater acute toxicity than that of glyphosate; it causes gastrointestinal damage, it affects the central nervous system, it causes respiratory problems, it destroys the red corpuscles, it damages the liver and the kidneys, and it causes skin injuries (erythema, swelling, exudation, ulceration), nausea, and diarrhea. In addition to that, the polyoxyethylene amine is polluted with 1-4 dioxane which has been associated with cancer development in animals and hepatic and kidney injury in human beings (Nivia, 2001).

The acute toxicity of the surfactant POEA is between 4 and 5 times greater than that of glyphosate and Roundup. The lethal oral dose (LD_{50}) in rats is > 1,200 mg/Kg and the dermal LD_{50} in rabbits is > 1,260 mg/Kg (CCE, 2007).

2.2. Cosmo-Flux 411F

There is little information on the composition and damaging effects of this chemical product. It has been used in the aerial spraying that took place in the Ecuadorian Northern Border as a nonionic coadjuvant that is added to herbicides like Roundup, in order to improve the adhesion and uniformity of the preparation of agrochemical substances. This compound is made up of List 3 products that are moderately toxic, and its average lifespan varies from 3 to 60 years (Cox, 1991 y 1995).

Its effectiveness is considered to be 4 times greater than the conventional spray oils due to the synergism between the paraffinic oil and the specialized tensioactive.

According to the technical card 313.03 of the Colombian company Cosmagro, which commercializes this product, Cosmo-Flux 411F is chemically described as a mix of mineral oil and specialized nonionic surfactants with coupling agents.

The active component is described as a mix of Hexitan esters; linear alcohols and ethoxylated aryl. In other words, it is a mix of stereospecific nonionic tensioactive based on linear ethoxylated propoxylated alcohols with small quantities of ethoxylated aryl (Cosmagro, 2004).

The additive component is made up of liquid isoparaffins that are highly pure isoparaffin oils with very low phytotoxicity, very low aromatic content, and low surface tension that improve humectability, thus promoting the efficiency of the active components (Cosmagro, 2004).

The Cosmo-Flux 411F is a yellow liquid at 77°F; flash point over 300.2°F; specific gravity of 0.89; viscosity of 60 mPa at 77°F; pH in solution at 1% in distilled water between 6.3 and 6.8, and oil concentration 80-85 (% w/w). It is soluble in mineral oils, vegetable oils, and organic solvents used in the formulation of agrochemical substances. It is dispensable in water generating a quick emulsion, and it is insoluble in ethanol (Nivia, 2001).

This stereospecific nonionic coadjuvant, made up of the combination of paraffin oil and a stereospecific tensioactive adjuvant, takes control of the evaporation of the active and the hydrolysis of the active with total covering, guaranteeing homogenous concentration of the active per area unit and expanding the spectrum of the agrochemical substances (Nivia, 2001).

2.3. N-Nitroso Glyphosate (NNG)

The N-Nitroso Glyphosate (NNG) has been recognized as a micropollutant of the technical active material and the concentrations found have been from 0.1 to 0.4 ppm (the detection limit has been from 0.1 to 0.2 ppm) (Onil, 2001).

When combined with nitrate (found in human saliva or fertilizers), glyphosate can generate traces of N-Nitroso Glyphosate. Most of the compounds of N-Nitroso are carcinogenic and there is no safe exposure level for them (Cox, 1995; Dinham, 1999).

3. Plant Genetic Resistance to Glyphosate

3.1. Naturally Acquired Resistance against Herbicides

Resistance is the natural inherent capacity that plants have for survival and reproduction after a treatment based on glyphosate. Plants, as such, are affected by the application of herbicides, but those that are naturally resistant survive and finish their reproductive cycle. Resistance to one or several herbicides is an inherited characteristic that may pass on to the following generations, depending on the nature of the involved genes, as well as the selection pressure imposed by the herbicide.

The resistant biotype is developed from a natural and spontaneous mutation that may be unique, and that may create a special individual among the species. This mutation has no relationship to the herbicide that is applied. This means that herbicides do not cause a change (mutation), creating the resistant biotype. Nevertheless, this biotype can widely spread itself throughout the population; that is, it is positively selected to the detriment of the sensitive individuals.

The evolution of naturally resistant plants to herbicides such as glyphosate leads over the years to serious ecological and environmental problems due to the adaptive capacity of the resistant plants and the probable extinction of the plants vulnerable to glyphosate. Since nature forms niches of different animal and vegetal species, the insertion of chemical substances like herbicides or pesticides alters this balanced natural chain, leading not only to the death of plants, but also to the death of thousands of insect species that live and feed on them.

Plant resistance to a certain herbicide takes places because of the appearance of one or a group of plants that provide naturally modified genetic information in a certain gene; and consequently, the specificity of herbicides in inhibiting certain enzymes codified by that gene does not happen. Thus, the plant is capable of being naturally selected for survival and reproduction in said environment.

By the end of 2005, 296 biotypes of plants resistant to herbicides were registered. Those biotypes belong to 178 different plants species: 107 dicotyledonous plants and 71 monocotyledonous plants. There are 90 plant species resistant to herbicides that inhibit the enzyme acetolactate synthase (ALS), 34 plant species resistant to herbicides that inhibit the enzyme acetyl-CoA carboxylase (ACCasa), and a minimum of 7 plant species resistant to glyphosate that inhibit the enzyme 5-enolpyruvylshikimate 3-phosphate synthase (EPSPS) have been confirmed.

According to the company Monsanto in Argentina, preventing the plants from being resistant to glyphosate entails the application of the adequate dose at the right time and with the adequate formulation of herbicide. In spite of the recommendations made by the company itself, the aerial spraying on the Ecuadorian border was carried out with a percentage of an active component and surfactant that is larger than the allowed amount without considering other important aspects such as genetic resistance, fauna, and human health.

3.2. Transgenic Plants Tolerant to Glyphosate and its Effects on Health

The soybean from Argentina, which is a genetically modified plant (transgenic) in order to acquire resistance to glyphosate, has been widely traded since 1996 for farming in that country (National Agriculture Statistics Service, 2005). Like the soybean from Argentina, the principle of transgenesis is being applied to several farming products worldwide.

The herbicides containing glyphosate as an active component, such as Roundup, work on the inhibition of the enzyme EPSPS on the shikimate pathway; this pathway is essential for plant life. The transgenic plants present a version of the enzyme EPSPS tolerant to glyphosate. The transgenesis process isolates the gene CP4 EPSPS of the *Agrobacterium* sp. strain CP4 resistant to glyphosate; this gene is cloned and once it is transfected to the soybean, this plant acquires the capacity to be tolerant to spraying with glyphosate due to the codification of the enzyme CP4 5- enolpyruvylshikimate 3-phosphate synthase tolerant to glyphosate (Heck et al., 2005; Funke et al., 2006).

In a report published in 2002 by the English *Institute of Science in Society* (ISS), Professor Joe Cummis reveals that the sanitary alert due to the presence of toxic acrylamide in cooked food is related to glyphosate, the herbicide that is tolerated by the most known transgenic crops, such as the Roundup Ready soybean. Acrylamide is a powerful neurotoxin that affects the male reproductive system and causes congenital malformations, and cancer in animals (Cummis, 2002).

The evidence and the studies that have been conducted show that acrylamide is released by the environmental polyacrylamide, the main source of which is found in glyphosate-based herbicide formulations. Cooking vegetables that have been exposed to glyphosate used in transgenic crops tolerant to herbicides or used in the preparation of land in conventional crops would result in an additional release of acrylamide. The situation is more complex when (in countries like the United States) the polyacrylamide-like additives are considered a "commercial secret" and the information on the composition of the herbicide formulations is not in the public domain (Smith et al., 1997; Weiss, 2002).

References

Acquavella, J; Alexander, B; Baker, B; Blleeke, M; Bruce, H; Champan, P; Gustin, C; Mandel, J. Glyphosate biomonitoring for farmers and their families: Results from the farm family exposure study. *Environ Health Perspect.*, 2004, 112, 321-326.

Agency for Toxic Substances and Disease Registry (ATSDR). ATSDR ToxProfiles 2002 on CD-ROM [On line]. 2001 [2010/08/10]. Available from: URL: www.atsdr.gov.

Budavari, S. *The Merck Index: An encyclopedia of chemicals, drugs, and biological.* 11th edition. Rahway, New Jersey: Merck & Co. 1989.

Chang, C; Hu, W; Hung, D; Lin, T; Peng, Y; Yang, D. Clinical impact of upper gastrointestinal tract injuries in glyphosate-surfactant oral intoxication. *Hum Exp Toxicol.*, 1999, 18(8), 475-478.

Comisión Científica Ecuatoriana (CCE). El sistema de aspersiones aéreas y su impacto en las condiciones de vida, salud y en el ambiente. *El sistema de aspersiones aéreas del plan Colombia y sus impactos sobre el ecosistema y la salud en la frontera ecuatoriana.*, 2007, 19-121.

Cosmoagro. Cosmo-Flux 411F, Coadyuvante de la aplicación de Agroquímicos. Lic. ICA 05.4-2186 - Colombia. Hoja Técnica 313.03 Mayo 30/94. Palmira, Colombia. 3. 2004.

Cox, C. Glyphosate fact sheet. *J Pestic Reform*., 1991, 11(2).

Cox, C. Glyphosate, part 2: Human exposure and ecological effects. *J Pestic Reform*, 1995, 15(4), 14-20.

Cummis, J. Acrylamide in cooked foods: The glyphosate connection [On line]. *The Institute of Sciences in Society (ISIS)*. 2002 [2011/10/03]. Available from: URL: http://www.i-sis.org.uk/acrylamide.php.

De Ruiter, H; Meinen, E. Influence of water stress and surfactant on the efficacy, absorption, and translocation of glyphosate. *Weed Sci*., 1998, 46(3), 289-296.

Denis, M; Delrot, S. Effects of salts and surfactants on foliar uptake and long distance transport of glyphosate. *Plant Physiol Biochem* (Paris). 1997, 35(4), 291-301.

Dinham, B. "Life sciences" take over. In: *Pesticides News*, 44, 7. The pesticides trust. PAN-Europe. London, UK. 1999.

Doliner, L. Emploiavantrécolte du glyphosate (RoundupMD). Document de Travail. *Agriculture Canada, Direction des pesticides*., 107, 1991.

Duke, S. & Powles, S. Glyphosate: A once-in-a-century herbicide. *Pestic Manage Sci*., 2008, 64, 319-325.

Feng, J; Thompson, D. Fate of glyphosate in a Canadian forest watershed: 2. Persistence in foliage and soils. *J Agric Food Chem*., 1990, 38(4), 1118-1125.

Funke, T; Han, H; Healy-Fried M, Fischer, M; Schönbrunn E. Molecular basis for the herbicide resistance of Roundup Ready crops. *PNAS*., 2006, 103(35), 13010-13015.

Garry, V; Burroughs, B; Tarone, R; Kesner, J. Herbicides and adjuvants: An evolving view. *Toxicol Indust Health*., 1999, 15(1-2), 159-167.

Gerritse, R; Beltran, J; Hernandez, F. Adsorption of atrazine, simazine, and glyphosate in soils of the Gnangara Mound, Western Australia. *Aust J Soil Res*., 1996, 34(4), 599-607.

Giesy, J; Dobson, S; Salomon, K. Ecotoxicological risk assessment for Roundup® herbicide. *Rev Environ Contam Toxicol*., 2000, 167, 35-120.

Glass, R. Adsorption of glyphosate by soils and clay minerals. *J Agric Food Chem.*, 1987, 35(4), 497-500.

Heck, G; Armstrong, C; Astwood, J; Behr, C; Bookout, J; Brown, S. Genomics, molecular genetics & biotechnology. *Crop Sci.*, 2005, 44, 329-339.

Kiely, T; Donaldson, D; Grube, A. Pesticides industry sales and usage 2000 and 2001 market estimates. *US Environmental Protection Agency, Office of Pesticide Programs (US EPA/OPP)*. Washington DC, USA. 2004.

Knisel, W; Davis, F; Leonard, R. GLEAMS Version 2.0 User Manual. *US Department of Agriculture, Agricultural Research Service, Southeast Watershed Research Laboratory. Tifton, GA*. 202. 1992.

Laerke, P; Streibig, J. Foliar absorption of some glyphosate formulations and their efficacy on plants. *Pest Sci.*, 1995, 44(2), 107-116.

Lin, N; Garry, VF. In vitro studies of cellular and molecular developmental toxicity of adjuvants, herbicides, and fungicides commonly used in Red River Valley, Minnesota. *J Toxicol Environ Health.*, 2000, 60(6), 423-439.

Miller, D; Griffin, J; Richard, E. Johnsongrass (Sorghum halepense), control and rainfastness, with glyphosate and adjuvants. *Weed Tech.*, 1998, 12(4), 617-622.

Mladinic, M; Berend, S; Vrodoljak, A; Kopjar, N; Radic, B; Zeljezic, D. Evaluation of genome damage and its relation to oxidative stress induced by glyphosate in human lymphocytes in vitro. *Environ Mol Mutag.*, 2009, 1-8.

National Agriculture Statistics Service. In Acreage eds. Johanns, M. &Wiyatt, SD. 6-30. US Department of Agriculture. Washington DC, USA. 2005.

Newton, M; Howard, K; Kelpsas, B; Danhaus, R; Lottman, C; Dubelman, S. Fate of glyphosate, in an Oregon (USA) forest ecosystem. *J Agric Food Chem.*, 1984, 32(5), 1144-1151.

Nivia, E. Cosmo-Flux 411F, coadyuvante adicionado, al Roundup Ultra en la erradicación forzosa, de cultivos ilícitos en Colombia [On line]. 2001 [2010/07/05]. Available from URL: www.usfumigation.org.

Onil, S. Riesgos inherentes a la utilización del Roundup, para el control de plantaciones de cocaína en Colombia. *Institut National de Santé Publique de Quebec*. 2001.

O'Sullivan, P; O'Donovan, J; Hamman, W. Influence of nonionic surfactants, ammonium sulfate, water quality and spray volume on the phytotoxicity of glyphosate. *Can J Plant Sci.*, 1981, 61(2), 391-400.

Piccolo, A; Celano, G; Arienzo, M; Mirabella, A. Adsorption and desorption of glyphosate in some European soils. *J Environ Sci Health Part B.*, 1994, 29(6), 1105-1115.

Reinert, K & Rodgers, J. Fate and persistence of aquatic herbicides. *Rev Environ Contam Toxicol.*, 1987, 98, 86-98.

Riechers, D; Wax, L; Liebl, R; Bullock, D. Surfactant effects on glyphosate efficacy. *Weed Tech.*, 1995, 9(2), 281-285.

Schuette, J. Environmental fate of glyphosate. *California EPA, Environmental Monitoring and Pest Management.*, 1998, 13.

Siltanen, H; Rosenberg, C; Raatikainen, M; Raatikainen, T. Triclopyr, glyphosate and phenoxyherbicide, residues in cowberries, bilberries and lichen. *Bull Environ Contam Toxicol.*, 1981, 27(5), 731-737.

Smith, A & Aubin, A. Degradation of 14C-glyphosate in Saskatchewan soils. *Bull Environ Contam Toxicol.*, 1993, 50(4), 499-505.

Smith, E & Oehme, F. The biological activity of glyphosate to plants and animals: A literature review. *Vet Hum Tox.*, 1992, 34, 531-543.

Smith, E; Prues, S; Ochme, F. Environmental degradation of polyacrylamides: II Effects of outdoor exposure. *Ecotoxicol Environ Saf.*, 1997, 37, 76-91.

Sundaram, A. Effect of a nalco-trol II on bioavailability of glyphosate in laboratory trials. *J Environ Sci Health Part B Pestic Food Contam Agric Wastes.*, 1990, 25(3), 309-332.

Sundaram, A; Leung, J; Webster, G; Nott, R; Curry, J; Sloane, L. Effect of glycerol on spreading and drying of vision droplets containing Silwet l-77: Relevance to rainfastness and herbicidal activity of glyphosate on trembling aspen Populustremuloidesmichx. *J Environ Sci Health Part B Pestic Food Contam Agric Wastes.*, 1996, 31(4), 901-912.

Thompson, D; Pitt, D; Buscarini, T; Staznik, B; Thomas, D; Kettela, E. Initial deposits and persistence of forest herbicide residues in sugar maple (Acer saccharum) foliage. *Can J Forest Res.*, 1994, 24(11), 2251-2262.

Tsui, M & Chu, L. Aquatic toxicity of glyphosate based formulations: Comparison between different organisms and the effect of environmental factors. *Chemosphere.*, 2003, 52, 1189-1197.

United States Drug Administration Forest Service. *Glyphosate: Human health and ecological risk assessment final report. USDA.* Virginia, USA. 2003; 39.

United States Environmental Protection Agency (US EPA). Description and sampling of contaminated soils. A field pocket guide. *Technology Transfer.* Ohio, USA. 1991.

United States Environmental Protection Agency, Office of Drinking Water (US EPA/ODW). Drinking water criteria document for glyphosate. US EPA. Office of Assistant Administrator. Washington, DC. 92-173392. 1992.

Weiss, G. Acrylamide in food: Uncharted territory. *Science.*, 2002, 297, 27.

World Health Organization (WHO). International Program on Chemical Safety (WHO IPCS). *Glyphosate.* Geneva., 1994, 159.

Zeneca to manufacture glyphosate, *Agrow.*, 1995, 230.

Chapter 2

Toxicology

Abstract

This chapter discusses toxicology and how different physical, chemical, and biological agents cause damage to different species in nature. Regarding genetic toxicology and its effects on health, the chapter refers to data published in several countries around the world where the population has been affected by toxicological agents. This chapter also deepens the analysis of the effects of pesticides both on human beings and other organisms, such as amphibians, rabbits, and sea urchins. As for human beings, the data obtained from research conducted in Ecuador as well as other countries affected by the excessive use of pesticides like glyphosate is detailed.

1. Genetic Toxicology and Its Effects on Health

Due to the industrial and agricultural revolution in the 20th century, as well as the application of new technologies for different purposes such as vegetable farming, chemical industry, and energy production, the presence of physical and chemical agents has been increasing.

Since 1940 it has been proven that the use of these chemical agents like pesticides can cause harmful effects on living beings, from bacteria to the most evolved mammals, including human beings.

Genetic toxicology essentially deals with the identification and analysis of the action carried out by any physical, chemical, or biological agent capable of causing toxic effects on human genetic material.

It is currently known that a good number of chemical products, both natural and synthetic, as well as ionizing radiation, UV light, some fibers and different viruses and bacteria can cause genetic alterations in cells and organisms. If the affected cells belong to the germ or reproductive line, they cause different genetically inherited defects. If the affected cells are somatic, mutations may lead to cancer, aging, and atherosclerosis. Likewise, it is also known that these potentially mutagenic agents may cause harm during the embryonic development, causing fairly serious alterations during normal development. Therefore, it is important to bear in mind that, although in most cases teratogenesis has no genetic base, there are genetic agents with teratogenic actions.

Teratogenic exposure commonly causes multiple anomalies, being responsible for around 10% of all congenital anomalies in humans. For instance, it is known that ionizing radiation can cause microcephaly; radioactive iodine can cause cretinism; polychlorinated biphenyls (PCBs) can cause fetal growth retardation, diffuse hyperpigmentation, and neonatal teeth; alcohol is responsible for a characteristic syndrome, including growth deficiency, mental retardation, behavior anomalies, and a typical facial aspect; cocaine has been related to breaking of the placenta and vascular breakage, such as fetal encephaloclastic lesions; thalidomide can cause different effects, especially serious anomalies in the growth of limbs during the first trimester of pregnancy; aminopterin and other cytotoxic compounds quickly destroy the fetal growing cells, causing growth deficiency and an array of anomalies.

Considering that the adverse consequences that the genotoxic agents cause to living beings and the relationship between genetic toxicity, mutation, cancer, and other diseases are important, genetic toxicology plays a relevant role in order to improve public health. Genetic toxicology has undoubtedly contributed to awareness of how important it is to preserve our genetic heritage and to fully reduce the risk that exposure to mutagenic and/or carcinogenic agents means to any human being (Johansen & Mange, 1998; Paz-y-Miño et al., 2002).

2. Toxicological Analyses of Pesticides in Humans

Glyphosate is one of the most used herbicides in farming worldwide, with 20,000 tons in Europe and 51,000 tons in the United States (Acquavella et al., 2004; Kiely et al., 2004). The activity of glyphosate is mainly based on the inhibition of the enzyme 5-enolpyruvylshikimate 3-phosphate synthase (EPSPS), generating delay of the metabolic route of shikimato, which is involved in the synthesis of aromatic amino acids in plants and microorganisms (USDA, 2003; Mladinic et al., 2009). It is commonly formulated with surfactants that decrease the surface tension of the solution and increase the infiltration of the tissues. Roundup is a water solution of isopropylamine salt of glyphosate, including POEA as a surfactant and CosmoFlux 411F as an adjuvant (Bolognesi et al., 2009; Tsui & Chu, 2003).

There have been several studies conducted worldwide that prove that the use of the formulation of glyphosate develops both high and low levels of toxicity in different organisms. This herbicide may interfere with some enzymatic functions in animals, but the poisoning symptoms depend on the dose and the exposure time.

As for human beings, it has been proven that glyphosate is toxic for placental and embryonic cells, as well as in the biosynthesis of sex steroids (Benachour & Séralini, 2009). This pesticide mixed with adjuvants has turned out to be cytotoxic by means of alterations of the succinate dehydrogenase enzyme which has also been proven to be toxic for the peripheral blood mononuclear cells (Martínez et al., 2007). Four case-control studies suggest an association between glyphosate and the risk of developing non-Hodgkin lymphoma (DeRoos et al., 2003; Eriksson et al., 2008; Hardell et al., 2002; McDuffie et al., 2001).

In 2007 a study related to the evaluation of the DNA damage in the Ecuadorian population exposed to glyphosate was published in the magazine *Genetics and Molecular Biology* (Paz-y-Miño et al., 2007).

The analysis of DNA fragmentation was carried out through the comet assay technique (The Single Cell Gel Electrophoresis assay), which is very useful when it comes to investigating DNA damage and its possible co-relationship with repair mechanisms. The comet assay has an excellent sensitivity and specificity for damage of the genetic material; when the cell nucleus is exposed to electrophoresis, the broken fragments move out of the cell nucleus, giving the image of a comet, depending on the size of the DNA fragments, as well as the quantity of the destroyed genetic material. The damaged cell has two options: 1) the DNA is repaired by its own system, 2) the DNA is not repaired for circumstantial or constitutional defects: if the germ cells are affected, reproductive difficulties or teratogenic effects on the offspring will be triggered. If the somatic cells are affected, cell apoptosis will take place. Or if the mutation cannot be eliminated, it will generate the cell line that causes cancer.

The objective of the analysis was to determine the supposed influence of the formulation of glyphosate used during the aerial spraying in 2006 on the genetic material of the exposed individuals. The exposed group consisted of 24 individuals who lived up to 1.9 miles from the border between Ecuador and Colombia. This group was confirmed to have toxicity symptoms after being exposed to the aerial spraying several times, such as: intestinal soreness and nausea, diarrhea, fever, headache, insomnia, tiredness, eye and skin burn, and respiratory problems (Paz-y-Miño et al., 2007).

Following two analyses, it was found that the individuals exposed to glyphosate showed high levels of migration of DNA in contrast to the control individuals ($p < 0.001$). The exposed individuals showed a mean migration of 35.50 μm, whereas the control individuals had a mean migration of 25.94 μm. The classes of comets D and E were not observed in the control group (Paz-y-Miño et al., 2007) (Figure 2.1) (Tables 2.1 and 2.2).

The compounds formulated from the mix of herbicides with glyphosate are more powerful mutagens for microorganisms, plants, animals, and humans, unlike the mechanism of action of the glyphosate itself. As a result, the individuals exposed to glyphosate had high levels of DNA damage in contrast to those who were not exposed to the aerial spraying.

In order to generate a space for analysis and academic and scientific thinking on the sanitary condition of the fumigated communities, as well as providing a space to listen and communicate with the members of the health teams who had been facing or avoiding this problem, the School of Medical Sciences at the *Universidad Nacional de Córdoba (UNC) de Argentina* called for the First National Medical Meeting of the Fumigated Communities to be held on August 27 and 28, 2010.

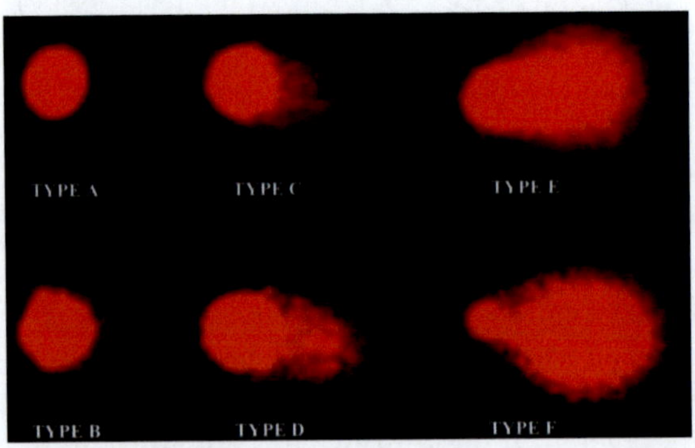

Figure 2.1. Classes of comets arranged by the level of DNA degradation Class A: Normal Nucleotide. Class E: Damaged Nucleotide.

Table 2.1. DNA damage measurements in individuals exposed to the aerial sprayings

Individuals exposed to glyphosate							
Individuals (sex, age)[a]	Number of cells analyzed in each group					Migration of DNA (μm)	
	A	B	C	D	E	Mean	Median
1E (F, 53)	2	120	76	5	3	39.5	32.5
2E (F, 37)	13	92	82	14	0	44.1	32.5
3E (F, 40)	2	64	62	77	4	56.6	52.5
4E (M, 27)	8	75	64	47	8	49.2	37.5
5E (F, 44)	9	138	63	3	0	34.6	30.0
6E (F, 50)	51	113	30	3	0	30.8	27.5
7E (F, 38)	21	139	48	3	0	33.2	30.0
8E (F, 46)	21	116	72	4	0	35.2	30.0
9E (F, 55)	26	100	84	1	0	32.8	30.0
10E (F, 50)	26	100	84	1	0	34.2	30.0
11E (F, 22)	28	123	60	0	0	32.0	27.5
12E (F, 27)	11	130	63	6	0	33.7	30.0
13E (F, 28)	40	132	40	2	0	31.0	30.0
14E (F, 59)	10	96	99	1	0	36.4	32.5
15E (F, 55)	35	110	62	1	0	32.7	30.0
16E (F, 17)	60	101	44	1	0	31.3	37.5
17E (F, 34)	7	114	57	2	0	33.4	30.0
18E (F, 45)	10	150	50	4	0	33.0	30.0
19E (F, 28)	13	160	44	0	0	31.3	27.5
20E (F, 21)	1	153	47	3	0	33.2	30.0
21E (F, 34)	2	130	25	1	0	31.8	30.0
22E (F, 23)	0	29	173	2	0	39.3	37.5
23E (F, 34)	2	88	115	1	0	35.5	37.5
24E (F, 42)	93	103	9	0	0	27.6	27.5
Mean Age = 38 ± 12.2[b]						35.5 ± 6.4[c]	30 ± 5.4[d]

[a]F = Female, M = Male; [b,c]Mean ± standard deviation; [d]Median value ± standard deviation

The expositions and stories from the participants coincided with the clinical observation of a range of diseases and conditions in the health of the population exposed to the fumigations. The most alarming thing for the doctors of the fumigated communities were two main remarks: 1) a larger number of newborns with congenital malformations and many more cases of spontaneous abortion than the usual number of cases were presented in their communities; and 2) more cases of cancer were detected in children and adults, as well as serious diseases, such as purpura, toxic hepatopathy, and neurological upset. The doctors highlighted the fact that they had attended to patients from the same communities for more than 25 years, but that what they had found over the last few was is totally unusual, and they linked this situation to the systematic fumigation with pesticides.

Table 2.2. DNA damage measurements in individuals who were not exposed to the aerial spraying

Individuals not exposed to glyphosate							
Individuals (sex, age)[a]	Number of cells analyzed in each group					Migration of DNA (μm)	
	A	B	C	D	E	Mean	Median
1U (F, 17)	150	59	3	0	0	26.2	25.0
2U (F, 40)	164	43	4	0	0	25.4	25.0
3U (F, 26)	165	40	2	0	0	25.7	25.0
4U (M, 14)	111	96	6	0	0	27.3	26.5
5U (M, 32)	165	38	3	0	0	25.9	25.0
6U (M, 21)	171	35	1	0	0	25.7	25.0
7U (M, 16)	177	25	6	0	0	25.8	25.0
8U (F, 47)	176	25	3	0	0	25.7	25.0
9U (F, 15)	190	14	1	0	0	25.2	25.0
10U (F, 36)	179	25	1	0	0	25.4	25.0
11U (F, 21)	150	46	9	0	0	26.3	25.0
12U (F, 43)	148	49	15	0	0	26.8	25.0
13U (F, 53)	161	27	10	0	0	26.1	25.0
14U (F, 35)	164	23	21	0	0	27.0	25.0
15U (F, 38)	169	28	11	0	0	26.4	25.0
16U (F, 22)	183	15	8	0	0	25.1	25.0
17U (F, 71)	191	8	5	0	0	25.0	25.0
18U (F, 39)	195	13	6	0	0	25.5	25.0
19U (F, 21)	179	20	8	0	0	25.9	25.0
20U (F, 50)	190	14	2	0	0	25.3	25.0
21U (F, 43)	150	56	9	0	0	26.4	25.0
Mean age = 33 ± 15[b]						25.94 ± 0.6[c]	25 ± 0.3[d]

[a]F = Female, M = Male; [b,c]Mean ± standard deviation; [d]Median value ± standard deviation.

Dr. Rodolfo Paramo, pediatrician and neonatologist at the Public Hospital in Malabrigo, Argentina, emphasized that it was alarming to learn about 12 cases of newborns with malformations among 200 births in 2006. This was in addition to 4 cases of children who ended up dead due to their birth congenital malformations in the small town of Rosario del Tala, an area where agrotoxic fumigation was extensive (PENMPF, 2010).

Dr. Maria del Carmen Seveso, chief of intensive therapy at the Hospital *4 de junio de Presidencia Roque Saenz Peña-Chaco*, presented a devastating situation in the central communities in the province of Chaco such as *Napenay, Gancedo, Santa Silvina, Tres Isletas, Colonia Elisa*, and *Avia Terai*, where there were many cases of people suffering from kidney disease, congenital malformations in young mothers' children, cancer in very young people, spontaneous abortion and difficulties in getting pregnant, and serious allergies and respiratory problems. The health teams linked all of these cases with a higher level of chemical pollution in the environment, generated by the agroindustrial practice imposed in the area which removed the small and numerous preexisting cotton lands and exterminated the native forest (PENMPF, 2010).

Dr. Ana Lia Otaño, national representative of the Ministry of Health in el Chaco, presented the results of an investigation which clearly highlighted an increase of the incidence of congenital malformations in newborns at provincial level, according to the data provided by the Neonatology Service at the *J.C. Perrando de Resistencia Chaco* Hospital (Table 2.3) (PENMPF, 2010).

In the province of Chaco in Argentina it is now known that the labor activity with agrochemical products or the exposure is related to reproductive problems, repetitive spontaneous abortions, and serious congenital malformations, such as the array of cases involving congenital malformations where the mothers have a history of direct exposure to pesticides, analyzed by Dr. Horacio Lucro, chief of the molecular biology lab at the Regional Medicinal Institute at the *Universidad Nacional del Nordeste*. He has been recording and studying the cases for more than 10 years; his remarks have been totally confirmed (PENMPF, 2010).

Regarding child cancer, Dr. Otaño pointed out that the incidence rate of cancer has markedly increased in comparison to the previously registered rates (Table 2.4).

Table 2.3. Increase of congenital malformations

Year	Number of Registered Malformation Cases	Alive Newborns	Incidence (malformed/ 10000 alive newborns)
1997	46	24030	19.1/10000
2001	60	21339	28.1/10000
2008	186	21808	85.3/10000

Table 2.4. Increase of child cancer cases

Year	Child Cancer Registered	Population under Age 15	Incidence
1985	29	275858	10.5/100000
1991	26	323788	8.03/100000
2001	40	354991	11.3/100000
2007	59	376833	15.7/100000

The incidence of cancer in the town most affected by agrotoxics (La Leonesa) was analyzed, and it was compared to moderately fumigated close towns (Las Palmas), and towns that were less fumigated (Puerto Bermejo) (Table 2.5) (PENMPF, 2010).

Table 2.5. Incidence of child cancer

Town	Population (2001)	Under Age 15 (2001)	Expected Cases	Registered Cases	Annual Incidence
La Leonesa	10067	2960	0.41 cases/ year	1996: 1 case 1997: 1 case 2000: 1 case 2003: 2 cases 2004: 1 case 2008: 1 case 2009: 1 case	1990 – 1999 0.2 cases/year 2000 – 2009 0.6 cases/year
Las Palmas	6593	2146	0.3 cases/ year	1993: 1 case 1995: 1 case 2006: 1 case	1990 – 1999 0.2 cases/year 2000 – 2009 0.1 cases/year
Puerto Bermejo	1832	652	0.09 cases/ year	1995: 1 case 2008: 1 case	1990 – 1999 0.1 cases/year 2000 – 2009 0.1 cases/year

The Latin American Congenital Malformations Center (ECLAM) reported that the Argentinean province of Misiones has a rate of 0.1/ 1000 alive newborns with neural tube defects. On the contrary, Dr. Demaio, pediatrician surgeon and chief of the Pediatrician Department at the *Posadas* Hospital in Misiones, recorded in his hospital a rate of 7.2/1000 alive newborns (70 more times). His team geographically indexed the origin of the families with these serious and disabling deficiencies and they all came from areas that were strongly fumigated (PENMPF, 2010).

Dr. Maria Fernanda Simoniello, together with the toxicology, pharmacology, and legal biochemical teams at the *Universidad Nacional del Litoral*, has focused on the analysis of biomarkers of cell reaction in people exposed to pesticides in a direct manner (fumigators in the fruit and vegetable industrial area in Santa Fe) and in an indirect manner (non-fumigators). These people have been evaluated through comet assay, in order to determine the DNA damage and fragmentation. This working group has several national and international publications (Simoniello et al., 2007, 2008, 2010).

The results showed that both groups of people exposed to pesticides had a rate of genetic damage which is considerably higher than the control group (not exposed to pesticides). This difference is statistically significant and it was kept in the reparation analysis of the genetic damage (Table 2.6).

On the contrary, there are studies that have proven a low level of glyphosate toxicity, such as the following: a study conducted by Bolognesi et al. in which a cytogenetic analysis of farmers from five regions in Colombia was made (Bolognesi et al., 2009); a study conducted by Sanin et al. in which it was proven that there is no association between glyphosate and the extended pregnancy period in women; and a study that proves the low glyphosate genotoxic risk in the environment, in contrast with products used in the cocaine production in Colombia (Solomon et al., 2009).

Table 2.6. Rate of DNA damage through comet assay

Comet Assay	**Control Group (n = 30)**	**Direct Exposure**	**Indirect Exposure**
Rate of damage (mean ± SD[a])	113.36 ± 13.48	214.92 ± 15.44	221.06 ± 18.32

[a]SD = standard deviation

3. Toxicological Studies of Pesticides in Other Organisms

Not only are there studies to evaluate the toxicity of a wide array of pesticides, such as glyphosate in human, but also research has been made on different animal species like amphibians, mice, rabbits, sea urchins, among others.

As for amphibians, *pipiens* Schreber frog tadpoles showed a decrease in the mouth-cloaca length and increase of the metamorphosis stage. Damage in the tail and anomalies at gonadal level were also found. The toxicity of pesticides is a factor that has contributed to the decreasing number of amphibians worldwide.

Regarding the development of the sea urchin eggs, glyphosate keeps enzymes from synthesizing and when the DNA is damaged, the CDK1/cycline B control point activates in the cell cycle, triggering the programmed cell death through apoptosis (Benachour et al., 2009).

Concerning rabbits, the treatment with glyphosate has caused body weight loss, low sperm concentration, and alteration in semen osmolality, whereas it has been observed in mouse liver mitochondrial samples that Roundup alters the mitochondrial complex II and III, allowing the formation of DNA adducts in both liver and kidneys.

The Molecular Embryology Lab of the National Scientific and Technical Research Center (CONICET-UBA), headed by Dr. Andres Carrasco, has been studying glyphosate as the agent of the embryonic development in vertebrates with recognized compatibility with the human embryonic development. Carrasco et al. verified the teratogenic effects of glyphosate, incubating and inoculating amphibian and chicken embryos with extremely diluted doses of herbicide. The obtained results were the following: decrease of the embryonic length, alterations in the formation of the embryonic axis, size modification of the cephalic region with effects in brain formation and eye reduction, alterations in the gill arches and the auditory placodes, and abnormal changes in the mechanisms of the neural plate formation that may affect the normal brain development, the neural tube closure, or other deficiencies in the nervous system (Paganelli et al., 2010).

When measuring the activity of some enzymatic systems, it was found that glyphosate increases the endogenous activity of the retinoic acid. The manifestation of the structural damage in embryos was reverted when the herbicide, an antagonist of the retinoic acid, was simultaneously used (Paganelli et al., 2010).

The authors conclude that the direct effect of glyphosate on the initial mechanisms of morphogenesis in vertebrate embryos causes concern due to the clinical results observed in the offspring of the population exposed to glyphosate in the agricultural land (Paganelli et al., 2010).

REFERENCES

Acquavella, J; Alexander, B; Baker, B; Bleeke, M; Bruce, H; Champan, P; Gustin, C; Mandel, J. Glyphosate biomonitoring for farmers and their families: Results from the farm family exposure study. *Environ Health Perspect.*, 2004, 112, 321-326.

Benachour, N; Séralini, GE. Glyphosate formulations induce apoptosis and necrosis in human umbilical, embryonic and placental cells. *Chem Res Toxicol.*, 2009, 22, 97-105.

Bolognesi, C; Carrasquilla, G; Marshall, E; Solomon, K; Volpi, S. Biomonitoring of genotoxic risk in agricultural workers from five Colombian regions: Association to occupational exposure to glyphosate. *J Toxicol Environ Health A.*, 2009, 72, 986-997.

De Roos, A; Zahm, S; Cantor, K; Weisenburger, D; Holmes, F; Burmeister, L. Integrative assessment of multiple pesticides as risk factors for non-Hodgkin lymphoma among men. *Occup Environ Med.*, 2003, 60, 1-9.

Eriksson, M; Hardell, L; Carlberg, M; Akerman, M. Pesticide exposure as risk factor for non-Hodgkin lymphoma including histopathological subgroup analysis. *Int J Cancer.*, 2008, 123, 1657-1663.

Hardell, L; Eriksson, M; Nordstrom, M. Exposure to pesticides as risk factor for non-Hodgkin´s lymphoma and hairy cell leukemia: Pooled analysis of two Swedish case-control studies. *Leuk Lymphoma.*, 2002, 43, 1043-1049.

Johansen, E & Mange, A. Basic Human Genetics. Sinauer Associates, Inc. Publishers Aunderland, Massachusetts. 1998.

Kiely, T; Donaldson, D; Grube, A. Pesticides industry sales and usage 2000 and 2001 market estimates. *US Environmental Protection Agency, Office of Pesticide Programs (US EPA/OPP).* Washington DC, USA. 2004.

Martínez, A; Reyes, I; Reyes, N. Cytotoxicity of the herbicide glyphosate in human peripheral blood mononuclear cells. *Biomedica.*, 2007, 27, 594-604.

McDuffie, H; Pahwa, P; McLaughlin, J; Spinelli, J; Fincham, S; Dosman, J. Non-Hodgkin´s lymphoma, and specific pesticide exposures in men: Cross-Canada study of pesticides and health. *Cancer Epidemiol Biomarkers Prev*., 2001, 10, 1155-1163.

Mladinic, M; Berend, S; Vrodoljak, A; Kopjar, N; Radic, B; Zeljezic, D. Evaluation of genome damage and its relation to oxidative stress induced by glyphosate in human lymphocytes in vitro. *Environ Mol Mutag*., 2009, 1-8.

Paganelli, A; Gnazzo, V; Acosta, H; Lopez, SL; Carrasco, AE. Glyphosate-Based Herbicides Produce Teratogenic Effects on Vertebrates by Impairing Retinoic Acid Signaling. *Chem Res Toxicol*., 2010, 23(10), 1586-1595.

Paz-y-Miño, C; Creu, A; Cabré, O; Leone, P. Genética Toxicológica y Carcinogénesis. AbyaYala. PUCE-FUNDACYT. Quito, Ecuador. 2002.

Paz-y-Miño, C; Sánchez, ME; Arévalo, M; Muñoz, MJ; Witte, T; Oleas, G; Leone, P. Evaluation of DNA damage in an Ecuadorian population exposed to glyphosate. *Genet Mol Biol*., 2007, 30(2), 456-460.

Primer Encuentro Nacional de Médicos, de Pueblos Fumigados (PENMPF). Facultad de Ciencias Médicas, Universidad Nacional de Córdoba (UNC). Córdoba, Argentina. 2010.

Simoniello, MF; Kleinsorge, EC; Scagnetti, JA; Grigolato, RA; Poletta, GL; Carballo, MA. DNA damage in workers occupationally exposed to pesticide mixtures. *J Appl Toxicol*., 2008, 28(8), 957-65.

Simoniello, MF; Kleinsorge, EC; Scagnetti, JA; Mastandrea, C; Grigolato, RA; Paonessa, AM; Carballo, MA. Biomarkers of cellular reaction to pesticide exposure in a rural population. Biomarkers., 2010, 15(1), 52-60.

Simoniello, MF; Scagnetti, JA; Kleinsorge, EC. Biomonitoreo de población rural, expuesta a plaguicidas. *Revista FACIBI*., 2007, 11, 73-85.

Solomon, KR; Marshall, EJ; Carrasquilla, G. Human health and environmental risks, from the use of glyphosate formulations to control the production of coca in Colombia: Overview and conclusions. *J Toxicol Environ Health A*., 2009, 72, 914-920.

Tsui, M & Chu, L. Aquatic toxicity of glyphosate based formulations: Comparison between different organisms and the effect of environmental factors. *Chemosphere*., 2003, 52, 1189-1197.

United States Drug Administration Forest Service. Glyphosate: Human health and ecological risk assessment final report. Virginia, USA, 39, 2003.

Chapter 3

AERIAL SPRAYING

ABSTRACT

The current chapter discusses the origin of the aerial spraying in Ecuador. The Amazon region is one of the most biodiverse areas in the world which obviously plays an essential role in the process of oxygen production. The Ecuadorian Northern bordering provinces, such as Sucumbios, Carchi, and Esmeraldas, have been exposed to the aerial spraying with glyphosate that was carried out during the period 2000-2007. Furthermore, this chapter deals with the characteristics of the aerial spraying with glyphosate, the perception the population obtained regarding this fumigation process, and the effects on different Amazonian communities, crops, animals, and water sources.

1. ECUADORIAN AMAZON BIODIVERSITY: RESOURCES AND THREATS

The Amazon region is one of the most important ecosystems or geographic regions on the planet and plays an essential ecological role worldwide. This region accounts for the third part of the total amount of tropical forests on the planet and its rivers account for the fifth part of fresh water flowing on the earth. The Amazon has the largest flora and fauna biodiversity. Likewise, it allows the regularization of oxygen and carbonic gas for South America and the world (Veyrunes, 2008).

Ecuador is a privileged country, having a part of this wonderful ecosystem and being home to the great biodiversity found in this area and other ecosystems in the Andean region. Therefore, Ecuador is one of the countries with the largest biodiversity worldwide with an area of only 104,480 square miles. This is due to the great variety of climate and geographic factors (Izquierdo et al., 2000).

In connection with the ecosystems found in Ecuador that go from the paramo at high altitude (up to 15,000 feet above sea level) to the cloud forest at middle altitude, the tropical humid forest at low altitude takes up a great area of the Ecuadorian territory, mainly in the Amazonian region. These forests are extremely important due to their area and ecological wealth, as well as their high biodiversity and heterogeneity, which are mostly reflected in the Amazon region (Palacios & Jaramillo, 2001).

The successive and final elevation of the Andean highlands caused the Western tropical vegetation to gradually become separated from the vast area that is currently known as the Amazon Basin. The weather turned more humid, generating an extraordinary flora with plenty of endemic elements (Van der Hammen, 2000). Approximately 80% of its endemic qualities and its biodiversity can be explained due to the prevalence of the allopatric speciation and the ecological and evolving balance in the region as a result of the elevation of the Andean highlands (Palacios & Jaramillo, 2001).

The Amazon region has often been described as a monotonous extension of canopy forest that covers more than half of the South American continent; nevertheless, there is high species heterogeneity in this ecosystem. The variation of species is found at basin, regional, and local level. After several studies, this heterogeneity has been proven by means of a limited distribution of vertebrates through rivers, which entails the presence of several endemic and sub-endemic species (Killeen, 2007). In Ecuador, in spite of its short extension in contrast with the total territory of the Amazon, the heterogeneity of this ecosystem is reflected in changing drivers of species dominance in the different provinces that are part of the Ecuadorian Amazon (Palacios & Jaramillo, 2001).

As for its biodiversity, the Ecuadorian Amazon is home to a great biological wealth, containing 27% of the tropical species and at least 13% of the plant species on the planet; consequently, Ecuador's Amazon tropical humid forest is one of the ten tropical biodiversity "hot spots". Additionally, Ecuador is an important place regarding amphibian wealth (third place worldwide with 402 species) (Izquierdo et al., 2000). Also, this is the tropical humid forest, mainly in the Amazon region, containing the largest percentages of amphibian and reptile species registered in the country (Izquierdo et al., 2000).

Considering the high diversity of the Amazon forest, the Amazon has an economic potential for the developing countries, such as Ecuador, and there are extraction activities of natural resources in a sustainable manner, destroying the ecological balance of this ecosystem, such as oil, mining, and wood extraction, as well as cattle raising and the use of pesticides in agriculture which are responsibly managed in some cases and irresponsibly managed in other cases; nevertheless, these activities always have an impact in the ecosystem.

In addition to the direct impact that these activities have in the ecosystem, side effects are generated, such as road construction in remote areas, which encourages humans to migrate, leading to a quick change of the use of the land and an increase in the fragmentation of the habitat. Since the land located at the base of the mountains and the contiguous prairies are mainly flat, the secondary roads multiply, causing huge deforestation, forest fragmentation and degradation (Killeen, 2007).

In Ecuador the annual deforestation rates vary from 0.5% to 2.4%; therefore, 46,000 and 220,800 forest hectares would be annually lost (Brito, 2006), leading to a decrease in the biodiversity and the alteration of the ecological balance in this fantastic ecosystem; life quality and the native Amazon communities are also affected.

In developing countries the plant biodiversity in tropical humid forests plays an important role for the people who live there, since some 80% of them use plant species as traditional medicine, which is an effective alternative to the treatment of many diseases. In Ecuador the traditional knowledge in the use of plants acquired and passed on through generations has evolved starting from the establishment of the first human beings up to now (Ríos et al., 2007).

The economic potential that the Amazon Basin is to Ecuador has prioritized the economic development over the care of this ecosystem; however, there are more sustainable alternatives to develop the wealth of its biodiversity. At world level it can be considered that 750,000 potential extracts can be synthesized from the plant biodiversity found in the tropical humid forest, in order to produce medicine for as much as $420 and $900 trillion dollars (Ríos et al., 2007).

The development in the research field regarding the medicine potential of the species biodiversity in the country has barely begun (Velasco, 2001), and may be a great alternative for generating income for the State; this way, the extraction activities of several resources may be replaced, mitigating the environmental damage and the loss of biodiversity.

2. ORIGIN OF THE AERIAL SPRAYING

Colombia is the country in which most coca (*Erythrozyllum coca*) grows worldwide. Likewise, it is one of the countries in which most opium poppies (*Papaver somniferum*) and marijuana plants (*Cannabis sativa*) grow (TAC, 2003). In order to eradicate the illegal crops, the Government of Colombia has promoted the strategy of fumigation in an aerial manner, by means of small airplanes and helicopters, with different high-concentration chemical herbicides (Crespo, 2009).

From 1978 to 1991 sporadic spraying with Paraquat, 2, 4-D and glyphosate was carried out in Sierra Nevada de Santa Marta (Puyana, 2007). From 1992 to 1999 systematic spraying took place. The National Narcotic Council (CNE in Spanish) regularized the proceeding and the spraying was granted the status of systematic public policy of State. In this period 187,858 hectares were sprayed, but drug production increased by 243%. During this period the Colombian Ombudsman's Office denounced the use of Tebuthiuron, Velpar (hexazinone), and Imazapyr (Castro, 2000; CCE, 2007; Puyana, 2007).

From the beginning, the aerial spraying was rejected by the affected population, by the political designers, and by the scientific community. In 1984 the Government of Colombia, by means of the National Hygiene Institute (INH in Spanish), gathered a group of experts in herbicides to analyze the potential damage caused by the aerial spraying. The experts were against the aerial fumigation with any herbicide, including glyphosate: "Its use in an aerial manner to eradicate marijuana and coca crops is not recommended". The data obtained in experiments with animals prove a low acute toxicity; its acute toxicity in human beings is very little known (TAC, 2003). In spite of the analysis and the recommendations provided by the experts, Colombia continued to carry out the aerial spraying with a concentrated mix of herbicides, in order to take control of the aforementioned illegal crops.

Colombia has depended considerably on the aerial spraying since 1999 when the *Plan Colombia* program between the United States and Colombia was established with the aim of eliminating the illegal crops through aerial spraying with different herbicides in the Colombian territory, including the areas bordering the Ecuadorian provinces of Sucumbios (Amazon), Carchi (Highlands), and Esmeraldas (Coast) (Crespo, 2009).

The first aerial spraying executed by the Government of Colombia took place in the provinces of Putumayo and Nariño, close to the Northern provinces of Ecuador. In connection with all the spraying activities in Colombia, it has been confirmed that several of them have affected the bordering Ecuadorian communities in a negative way. In the following table the different communities affected from 2000 to 2007 are observed.

3. History of the Impacts Caused by the Aerial Spraying

3.1. Characteristics of the Aerial Spraying

Following the aerial spraying carried out in connection with the *Plan Colombia*, 32.37% of the families affirmed that a relative became seriously sick while the crops and the animals were affected.

Table 3.1. Ecuadorian communities affected by the Colombian aerial spraying

Province	Community	Date
Esmeraldas	Mataje	October 2000
Carchi	San Marcos	October 2000
Sucumbios	San Francisco 1 and 2, Nuevo Mundo, and San Pedro del Condor	December – February 2001
Carchi	San Marcos	October – November 2001
Sucumbios	Chone II, Playera Oriental, Palma Seca, Puerto Nuevo, Santa Marianita, 5 de Agosto, and Puerto Mestanza	August – October 2002
Sucumbios	Santa Marianita, Corazon Orense, 5 de Agosto, and Puerto Mestanza	July 2003
Carchi	Chical	December 2004
Sucumbios	Frente al Azul	December 2004
Esmeraldas	Limones	April 2005
Carchi	San Marcos	May 2005
Sucumbios	Comunidades de Salinas a Puerto Escondido	December 2006
Sucumbios	Comunidades de Puerto El Carmen a Río Abajo	January 2007

The aerial fumigation that affected the bordering area with Ecuador was first noticeable in October 2000 in the province of Esmeraldas where the population in Mataje announced serious health consequences. In the area of Sucumbios the last period of fumigation was from December 2006 to March 2007. 57.64% of the population recognized this fumigation as the last activity that they experienced and that most of the activities took place in the Christmas season, with those of 2002, 2003, and 2004 being the most remembered sprayings because the people were affected several times. The sprayings, according to the family surveys, lasted from 1 to 20 days, including a 4-day break.

The number of spraying airplanes was between 1 and 3, escorted by 1 to 4 helicopters or sometimes more than that, for protection. 74.58% of those surveyed affirmed that the aerial spraying took place from the morning to the afternoon, 6 hours a day.

Even though half of the surveyed families pointed out that the sprayings were not different from one another, some 35.7% said the opposite.

As a matter of fact, the families mentioned that the amount of applied chemicals was different (18.92%), considering the chemical composition or the intensity of the impacts. Sometimes the difference of the chemical composition was visible (18.92%) due to their different colors.

3.2. Perceptions of the Aerial Spraying

The odor of the aerial spraying was widely defined. 46% of the surveyed individuals defined the odor as being like "poison", "chemical product", or "glyphosate", whereas 29.8% gave other options: "do not know", "sharp", "burned rubber", "powder", "acid", "fuel."

The visual aspect of the aerial spraying was also described extremely differently by the people surveyed. 39% affirmed the appearance was like fog, 13.48% said it looked like a "drizzle" or "mist", 19.15% said it was like "smoke" or "gas", and finally 21.7% did not witness the aerial spraying, so they had no opinion.

Regarding the color of the aerial spraying, 69.29% said it was "white" or "gray", 7.14% pointed out it was "black", and 2.8% affirmed it was "transparent" or "blue", whereas the remaining 20.07% gave other answers: "do not know", "did not see", "do not live here", or "the water looked white and the gas was black."

As for the consistency of the aerial spraying, 45.4% affirmed that it was oily and 33.3% claimed it was liquid. All these variations have supported concerns about the fact that the aerial spraying was not carried out with one single product, but several chemicals (Figure 3.1).

Among the families that use pesticides, 78% affirmed that the effect of the aerial spraying was very different from the normal use of chemicals in agriculture.

3.3. Behavior in View of the Aerial Spraying

The lack of information on the possible risks that the mix of pesticides may generate from the authorities responsible for the aerial spraying caused the population to feel insecure.

84.33% affirmed they had been carrying out their normal duties at home, on the farm, or doing the laundry in the river when the aerial spraying took place.

Watching the aerial spraying directly caused some 37.7% of the families to shut themselves in their houses, which did not guarantee their safety, since the houses are not hermetic due to weather conditions. In fact, 46.88% of the surveyed people affirmed that the chemicals had penetrated their houses, whereas 44.53% believed they had not.

Some 47.7% mentioned that they had no protection to continue with their duties and only 4.6% reacted to the situation; for instance, by immediately taking a shower, seeking protection under a tree; measures that prove that the population was defenseless against an event that they had not been informed was going to happen.

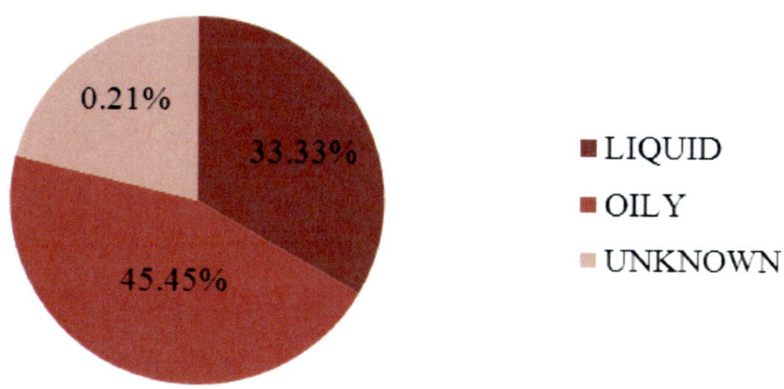

Figure 3.1. Consistency of the aerial spraying.

In Figures 3.2 and 3.3 the different reactions to the aerial spraying given by those surveyed and their families can be observed.

Figures 3.2 and 3.3 show the population's reaction to the aerial spraying. A combination of fear and curiosity was found among the people. Some shut themselves in their houses when there was a military deployment, while other families watched what happened from their hiding places. The reactions varied from curiosity at the beginning, to ignorance afterwards, with some even thinking that it may be beneficial. But later, the reactions changed.

Those who stated that they were afraid claimed to be concerned about a supposed "war", "an armed conflict", and that "they were going to die", since "there were gunshots at the small airplanes up the mountain in Colombia", but on the other hand, they were afraid of "the poison" and its effect.

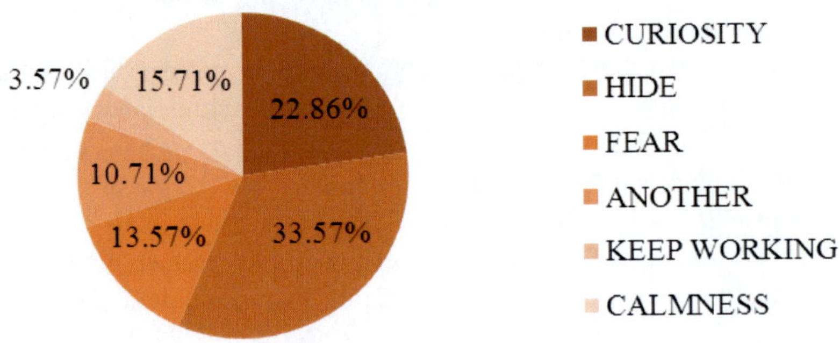

Figure 3.2. Survey results of respondents' reactions to the aerial spraying.

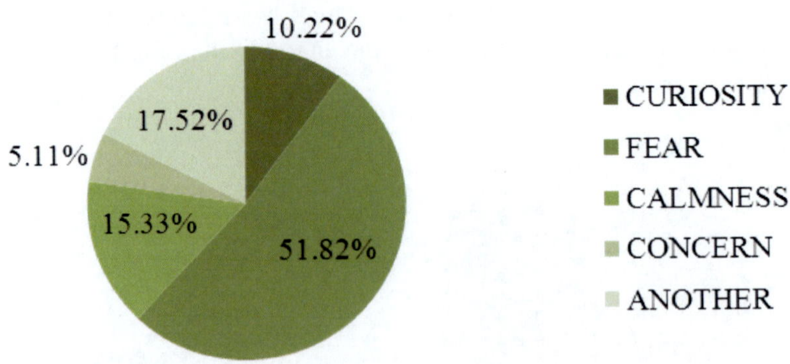

Figure 3.3. Families' reactions to the aerial spraying.

4. Effects of the Aerial Spraying on the Community

4.1. Community Development

During all the community assemblies it was said that the aerial spraying had a negative impact on the community development and so they discussed the impacts on health, feeding, and the diseases in children; every case on which it was necessary to spend economic resources. Nevertheless, this situation was worsened by the fact that the income decreased because the crops were reduced to almost nothing because the land was affected, so it became impossible to make it productive again. Therefore, the communities are still suffering from the impact of the aerial spraying. In some communities it was reported that the water was affected, as well as education due to the lack of teachers because they did not want to go to the bordering provinces. It was also mentioned that the coffee crops were lost along with other products that allowed the community to survive, keeping the families from commercializing their products and causing them to have low income. In one of the communities it was pointed out that animals had died.

4.2. Family Cohesion and Migration

During the community assemblies most of the communities by consensus affirmed that the aerial spraying forced them to stick together in order to better denounce the attacks they were undergoing, including the impact on their crops. Nevertheless, one of the communities commented that the aerial spraying caused families to fall apart, since many of them migrated at that time, although some are now coming back. In fact, in 9 out of the 10 communities it was mentioned that 49 families had left the area, which accounts for some 15.96% of families that had to move out.

The families that suffered from internal migration left their farms reassembled, hoping to come back. The population affirmed that it was not a huge migration because they had no place to go, and that the governmental support was devastating.

While part of the Ecuadorian population moved out of the affected areas, 45 Colombian families settled down at this time, accounting for some 14.65% of the families that currently live in the area. Some of these families were Paeces Indigenous.

The responses to the death of animals, the loss of crops, and family diseases were always made through individual efforts. Medicine was sought through veterinarians, families resorted to doctors or home medicine, and more than a single farm was harvested in order to face the problem. The only collective measure was to join the Association of Peasant Organizations from the Bordering Area (FORCCOFES) and to file a lawsuit at international level.

5. Effects of the Aerial Spraying on Crops

During the community assemblies the population commented that the aerial spraying on the short-cycle crops caused plants to become dried and die, and the grain plants became spoiled without a chance to produce, just like what happened with rice and corn. The sprouts of the coffee plant died and became yellowed until the seed and the leaves fell down and the plant could no longer bear fruit. In some communities production decreased from 30 quintals to fewer than 10 per hectare. As for cacao, it was found that the top of the trees had died, causing the leaves to become dried and yellow, and eventually to die, turning black. Concerning banana, they mentioned that the leaves of the plants turned yellow, they did not bear fruit, and the plants were black inside.

Regarding yucca, they said that these trees got spoiled; they were black inside as the leaves fell down, and the roots did not grow. As for the fruit trees, it was mentioned that they did not bear as much as before and the leaves, as well as the fruits, fell down.

In the family surveys it was described how plants were perceived during the aerial spraying: oily or sticky (22.35%), yellowish (22.35%), withered and dried (23.6%), burned (8.2%), and leaves and fruit falling down (3.6%). Some families argued that the trees had a bad smell (2.4%), became withered (2.4%) or blackish and became spoiled (2.4%) until they died (7.1%). 3.5% affirmed that plants were wet and 2.4% found everything normal.

A few days after the fumigation, these plants were described thus: yellowish (11.5%); the flowers, the fruit, and the leaves fell down (8.3%); they became withered and dried (28.1%); and they died (19.4%). Some families said that rather than getting dried, plants became spoiled (10.6%) and turned black (4%), as well as burned (5.7%). Other families pointed out that plants became weaker and were filled with fungus and plagues (2.6%), that they stopped growing up (1.7%), that they became sterile and stopped producing grains (5.7%), or that the leaves got broken, and turned spotted, hard, or damaged (2.1%). To summarize, 96.32% of the families commented that the aerial spraying caused plants and trees to have a different color, appearance, or flavor.

From one lot to almost 50 hectares, those surveyed mentioned that they saw how 1,141 hectares were damaged out of the 2,500 they collectively owned, accounting for 45.6% of their farms.

6. Effects of the Aerial Spraying on Animals

37.98% of the surveyed families commented that they had found dead fish in the rivers and estuaries; 59.69% did not see dead fish, and 2.33% did not know what to answer. Of the surveyed population, 19.6% affirmed they had pools with fish, of which each family lost an average of 467 fish.

As for cows, in the community assemblies the population affirmed that after the aerial spraying took place, the cows aborted in 5 communities; the calves stopped suckling, lost weight, and died, whereas the adult cows got swollen, had diarrhea, shivered, and some died. It was said that more than 13 cows died in the community of *Corazon Orense*.

Concerning pigs, it was mentioned that the babies died 2 days after the fumigation, while the adult ones died 7-10 days after the spraying. In one of the affected communities 3 families reported the loss of 270 chickens due to apparently polluted water. Dogs and cats lost their fur and had skin disease. This condition occurred in 6 out of the 10 studied communities.

The animals presumably affected by the aerial spraying were consumed by the same community.

The surveys made it possible to find out that 30.88% of cows, 35% of horses, 33.3% of chickens, and 33.8% of pigs died in 2002, followed by 2003 and 2001 as the years with the greatest losses.

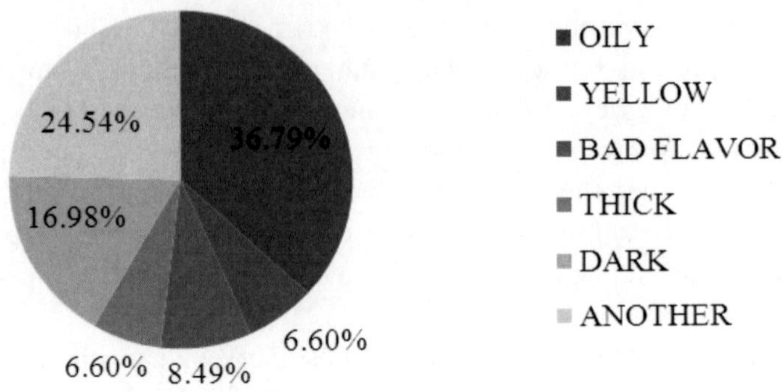

Figure 3.4. Aspects of the damaged water.

7. Effects of the Aerial Spraying on Water

68.9% of the surveyed families noticed that the water changed after the aerial spraying, while 31.1% did not find any change. Among the first group, 58.5% saw the water had turned oily or cloudy. 29.25% noticed the water had a different color, and 13.2% perceived a bad smell or flavor (Figure 3.4).

References

Brito, NF. Representaciones de la naturaleza en la Amazonia ecuatoriana: ¿Subsistencia local o conservación global?. *Iconos Rev Cie Soc.*, 2006, 25, 57-65.

Castro, JF. Los cultivos ilícitos, política mundial y realidad en Colombia. *Defensoría del Pueblo de Colombia.* Bogotá, Colombia. 2000.

Comisión Científica Ecuatoriana, (CCE). El sistema de aspersiones aéreas y su impacto en las condiciones de vida, salud y en el ambiente. *El sistema de aspersiones aéreas del plan Colombia y sus impactos sobre el ecosistema y la salud en la frontera ecuatoriana.*, 2007, 19-121.

Crespo, R. Demanda de la República del Ecuador. Instrumentos Internacionales de Derecho Ambiental. *Editorial de la Universidad Técnica Particular de Loja. Universidad Técnica Particular de Loja (UTPL).* Loja, Ecuador. 2009.

Izquierdo, J; Nogales, F; Yánez, AP. Análisis herpetofaunístico de un bosque húmedo tropical en la Amazonia ecuatoriana. *Ecotrópicos.*, 2000, 13(1), 29-42.

Killeen, TJ. Una tormenta perfecta en la Amazonia: Desarrollo y conservación en el contexto de la iniciativa para la integración de la infraestructura regional sudamericana (IIRSA). *Conserv Int.*, 2007, 48-49.

Palacios, WA; Jaramillo, N. Riqueza florística y forestal de los bosques tropicales húmedos del Ecuador e implicaciones para su manejo. *Rev Forest Centroamer.*, 2001, 36, 46-50.

Puyana, AM. Fumigaciones químicas como estrategia antinarcóticos en Colombia: tres décadas de perversidad e ineficacia. *Ponencia de la autora en la pre-audiencia sobre biodiversidad, fumigaciones y militarización en Colombia.* Medellin, Colombia. 2007.

Ríos, M; Koziol, M; Borgtoft, PH; Granda, G. Los etnobotánicos en el Ecuador: Escenario actual y desafíos. Plantas útiles del Ecuador: Aplicaciones, retos y perspectivas. *Ediciones Abya-Yala.* Quito, Ecuador. 652, 2007.

Tribunal Administrativo de Cundinamarca (TAC), Colombia. Segunda sección. Subsección "B*". "Claudia Sanpedro y Héctor Suárez contra Ministerio del Ambiente y otros*", 15, 2003.

Van der Hammen, T. Aspectos de historia y ecología de la biodiversidad norandina y amazónica. *RevAcadColombCienc.*, 2000, 24(91), 231-245.

Velasco, AM. Propuesta de Ecuador para la formulación de la estrategia nacional de biodiversidad: Vida silvestre, estudio nacional. *Estrategia regional de biodiversidad para los países del trópico andino.*, Quito, Ecuador. 2001.

Veyrunes, E. Las amenazas percibidas para la Amazonia: un estado del arte en términos de seguridad ambiental. Universidad del Rosario. Bogotá, Colombia. 2008, 7-8.

Chapter 4

Biomonitoring in the Province of Sucumbios

Abstract

The current chapter deals with the social, environmental, and health analyses carried out in the different communities located in the province of Sucumbios which borders Colombia. In addition to that information, the origin and the data of the studied population, the socio-economic indicators of the communities, the ethnic origin, age, sex, residence time, and the educational level of the population are explained in detail. Regarding life conditions, an analysis is made, focusing on the structure of the houses, the presence of windows and mosquito nets in the houses, trash management, water sources for human consumption, feces management, and land ownership.

1. Area of Study

The province of Sucumbios is located in the Amazon Region, Northeastern Ecuador. This province was created on February 13, 1989 due to the need for the Government of Ecuador to improve the administrative and economic system both in this province and in the country. Some of the most relevant goals that were set for the creation of Sucumbios are the following: a) to take advantage of the huge oil and mining wealth; b) to take control of the population pressure in a more efficient manner; and c) to meet the urgent basic needs of the population.

The jurisdictional reorganization of the province of Sucumbios made it possible to form seven cantons: Lago Agrio, Cuyabeno, Shushufindi, Sucumbios, Putumayo, Gonzalo Pizarro, and Cascales.

1.1. Communities

Socio-economic, genetic, and health analyses were conducted in ten communities located along the Ecuadorian border with Colombia, in the province of Sucumbios. The following communities were analyzed: Chone 2, Yanamarum, Playera Oriental, Fuerzas Unidas, Puerto Escondido, Corazon Orense, Santa Marianita, San Francisco, Las Salinas, and 5 de Agosto (Figure 4.1).

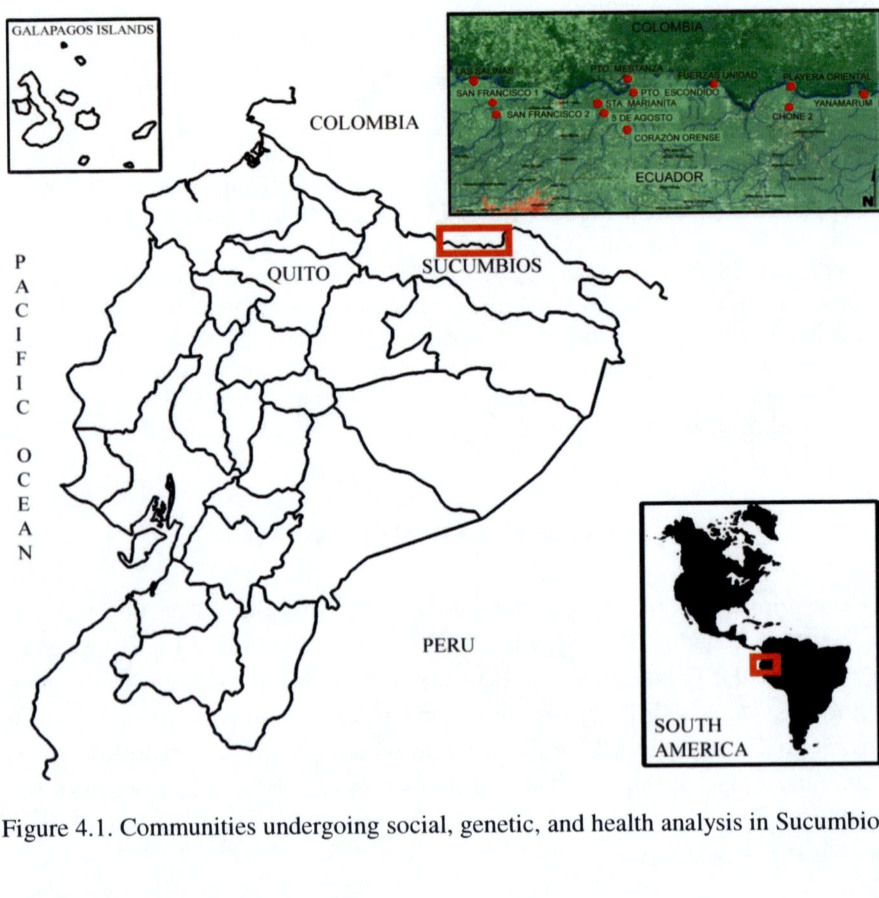

Figure 4.1. Communities undergoing social, genetic, and health analysis in Sucumbios.

These communities are located 1 mile away from San Miguel River. According to the population data, there are 307 families and 1,450 inhabitants, and they were selected after having previously announced that they had suffered from impacts due to the aerial spraying in connection with the *Plan Colombia*, after being as close as possible from the bordering line, and after having been previously analyzed.

In the ten communities that were selected due to their closeness to the border line with Colombia and after having been affected by the aerial spraying with glyphosate from the beginning of the *Plan Colombia*, specific areas at schools, health centers, and meeting rooms were chosen in order to make the respective interviews, to carry out medical examinations, and to take peripheral blood samples.

One of the facts of interest is that these communities have been previously evaluated in such a way that the medical and genetic results have been compared in order to get to know the general conditions of people's health. Data like UTM coordinates and altitude were collected in each of the sampling points (Table 4.1).

Environmental analysis was made in eighteen communities located by the San Miguel and Putumayo rivers in the bordering area with Colombia that belongs to the cantons Cascales, Lago Agrio, and Putumayo in the province of Sucumbios, with a tour extension of almost 295 square miles. The San Miguel River flows through the coordinates 00° 23´ 17´´ N 77° 10´ 60´´ O and 00° 07´ 54´´ N 76° 04´ 05´´ O, and the Putumayo River is located in the coordinates 00° 02´ 56´´ N 76° 24´ 00´´ O and 00° 07´ 27´´ N 75° 51´ 98´´ O.

The selection of the intervention zone was made by applying a statistical analysis of stratification, considering communities affected by aerial exposure to glyphosate during the period 2000 to 2007 in a minimum range of 0.1 mile and maximum range of 15 miles from the border.

The field phase planning consisted of four areas of study in which the following aspects were considered: physical, weather, physiological, geological, demographical, geographical, sanitary-environmental, community interest, and participation of the community health promoters.

2. Origin and Data of the Population

Of 307 registered families, 144 families (46.9%) comprising in total 771 people (53.17%) were surveyed. Of these families, a medical examination was conducted on 521 people (accounting for 35.93% of the people who live in the communities and 67.6% of the 771 people who are part of those families) (Table 4.2).

2.1. Socio-Economic and Health Indicators

Family surveys for one or two family heads were conducted. This information was useful for providing socio-economic and health indicators of the communities that were analyzed. Four types of history were applied: 1) community history, 2) family history, 3) personal clinical history, and 4) dermatological clinical history.

The community history consisted of obtaining information on the location of the community; its characteristics, such as ethnic group; the length of time since it was created; origin of the population and presence of oil activity; community cohesion; information on the mobilization; loss of territory and its consequences; health: types of diseases acquired by people, animals, and plants; levels of violence or protection on the part of the military groups; and governmental support like health center service, power service, public schools, and community laundry service.

Table 4.1. UTM Coordinates and altitude of the visited bordering communities

Province	Communities	Symbol	UTM Coordinates	Altitude (feet above sea level)
Sucumbios	Chone 2	CH2	18 N 331387 25220	967.8
Sucumbios	Yanamarum	YAN	18 N 342418 26190	830.1
Sucumbios	Playera Oriental	PLO	18 N 342418 26190	826.77
Sucumbios	Fuerzas Unidas	FUN	18 N 319301 28533	839.9
Sucumbios	Puerto Escondido	PTE	18 N 304907 27303	1000.66
Sucumbios	Corazon Orense	COR	18 N 305456 25473	902.23
Sucumbios	Santa Marianita	SMA	18 N 298861 25797	931.76
Sucumbios	San Francisco	SF2	18 N 286061 25056	261.29
Sucumbios	Las Salinas	SAL	18 N 281286 29627	1007.22
Sucumbios	5 de Agosto	5AG	18 N 301318 25497	1568.24

Table 4.2. Data of the analyzed families per community

Community	Registered Families	Analyzed Families	% Families	% Total	Population	Examination	% Com.	% Total
Chone 2	26	15	57.7	10.4	130	41	31.53	7.87
Yanamarum	13	9	69.2	6.3	65	48	73.84	9.21
Playera Oriental	17	9	64.7	7.6	86	46	53.49	8.83
Fuerzas Unidas	30	16	53.3	11.1	119	73	61.34	14
Puerto Escondido	18	14	77.8	9.7	120	44	36.67	8.44
Corazón Orense	40	18	45	12.5	200	46	23	8.83
5 de Agosto	28	12	42.8	8.3	140	43	30.71	8.25
Santa Marianita	60	20	33.3	13.9	300	62	20.67	11.9
San Francisco	22	13	59.1	9.0	110	67	60.9	12.8
Las Salinas	53	16	30.18	11.1	180	51	28.33	9.79
Total	307	144	46.9	100	1450	521	35.93	100

Concerning the family history, the data that was obtained focused on housing conditions like the house building material, number of rooms, beds, outlets, and the feces and trash treatment; regarding occupation and exposition, the number of hectares per farm, the type of crops and the use of agrotoxics were reported. As for the health situation, the numbers and the causes of dead relatives, the numbers of abortions and newborns with malformations, and the numbers of children who had passed away after the aerial spraying with glyphosate were collected. In regards to the aerial spraying, evidence of the duration, the distance that the airplanes flew over the houses, the physical and psychological effects during and after the spraying, the damage caused to land, plants, water, air, and the number of farm animals that had passed away after being in contact with glyphosate was collected.

Regarding the personal clinical history, the problems in organs and systems were evaluated. Information on problems found in stomach, eyes, ears, skin, hair, kidneys, urinary tract, nose, throat, nails, ganglions, teeth, respiratory system, circulatory system, bones, articulations, blood, endocrine system, genitals, breasts, and central nervous system was obtained. As for the background data, the survey sought information on intoxications, infections, prescribed medicine, and smoking and alcohol habits.

The dermatological clinical history made it possible to evaluate skin alterations and problems in the population. In this survey information on the presence of dermal or epidermal inflammation, necrosis, skin color, tumors, and hair and nail alterations was obtained.

2.2. Origin

53.4% of the population were born in the Amazon provinces, 19.8% come from the Highlands, 16.1% is from Colombia, and 10.7% come from the provinces in the Ecuadorian Coast. The analyzed population came from 17 Ecuadorian provinces except for Morona Santiago, Azuay, Cañar, Tungurahua, Cotopaxi, Santa Elena, and Galapagos. The distribution of the provinces is detailed in the upcoming table.

54% of the people who were born in Sucumbios had been born in the same studied communities with important differences that can be observed in Figure 4.2.

Table 4.3. Origin of the analyzed population by province

Origin by provinces							
Amazon	**Sucumbios**	**Napo**	**Orellana**	**Pastaza**	**Zamora**	**-**	**TOTAL**
	48.74%	1.94%	1.36%	0.77%	0.58%	-	53.4%
High-lands	Loja	Bolívar	Carchi	Chim-borazo	Imbabura	Pichincha	
	10.7%	4.08%	3.49%	0.97%	0.38%	0.2%	19.8%
Coast	Sto. Domingo	Manabí	Esmeraldas	El Oro	Los Ríos	Guayas	
	3.69%	2.91%	1.94%	1.16%	0.58%	0.38	10.7%

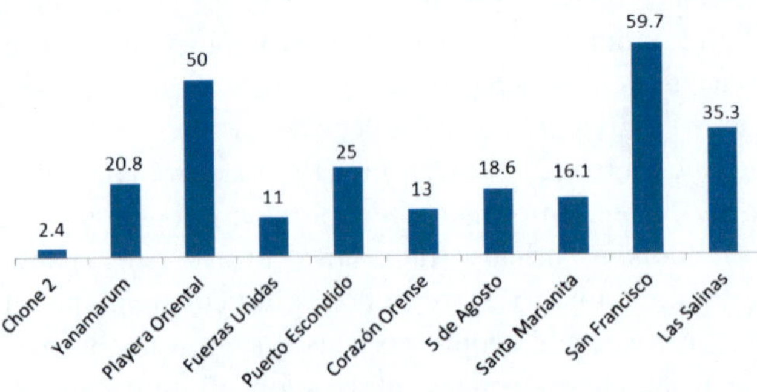

Figure 4.2. Percentage of births in the analyzed communities.

This data shows the communities in which women mostly give birth at home. 25% have always lived in the visited community. 29.6% of those who came from Sucumbios to the bordering areas live more than 20 years.

2.3. Ethnic Origin

Concerning the ethnic origin of the population, it was found that 83% of them are mestizo, 15% have an Indigenous origin, and approximately 2% are black. In these communities 14.1% of the people have Colombian origin, whereas the 85.9% have Kichwa origin (Figure 4.3).

2.4. Time of Residence

Regarding the studied communities, 13% have lived there more than 20 years and 27% from 11 to 20 years, the largest group being those who have lived in the communities in Sucumbios fewer than 10 years (60%). The mobilization of the last group is highlighted due to the fact that it coincides with the aerial spraying and if the 60% is taken as universal, it is found that almost 75% of the population who live in the border areas came from Colombia less than a decade ago, which raises the possibility that the reason for their moving out was the fumigation policy in the border that started with *Plan Colombia* in 2000. As for the last 5 years, 51.6% of the families that settled down are Colombian and 12.5% of the families arrived in the Ecuadorian border area after the aerial spraying; that is, 2 years ago.

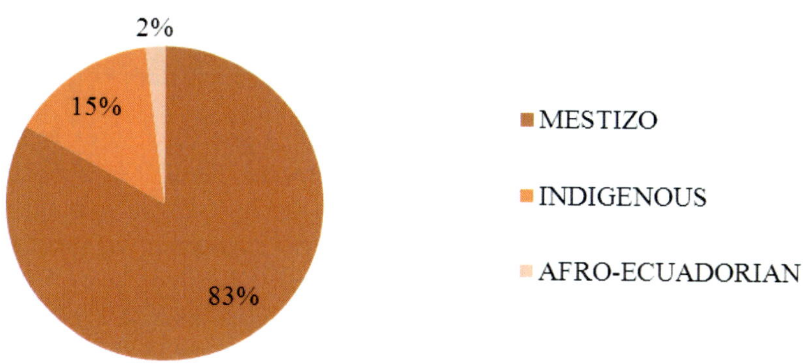

Figure 4.3. Percentage of groups of different ethnic origin in the analyzed population.

2.5. Sex and Age

In regards to sex, 47.8% of the studied people were men, in contrast with 52.2% women. The pyramid cut goes in accordance with a "pyramid of young population" characterized, according to INEC, by high fertility and even higher mortality (INEC, 2001).

More than 40% of the people are younger than 15 years old and less than 4% are older than 65 years old. In this study 49.13% of the surveyed population are younger than 15 years old and 1.34% are older than 65 years old. However, the continuous absences among the young male students make us think that it might be about migration processes.

2.6. Educational Level

In the studied communities the illiteracy level was 10.2% (with 13.2% men and 7.7% women). This data agrees with other analyses in which illiteracy data in individuals older than 18 years old is reported to be some 10.2% (Maldonado et al., 2008). It is the women who have a larger percentage of complete high school education (7.7%), unlike men (4.7%).

3. Life Conditions

It is important to get to know the communities' life conditions, so as to evaluate the socio-economic levels.

3.1. Distance to the Border Line

The average distance between the border line and the houses is around 0.78 miles, whereas the average distance of the farm is of 0.71 miles away. The surveyed family group is distributed in a range of less than 0.06 miles and up to more than 2.50 miles away. There is a group of surveyed people who have properties located between 0.06 miles and 0.25 miles away from the border line, and other surveyed people had properties located between 0.62 miles and 2.49 miles away. The results of the study averaged the situations and impacts that happened in the first front from the river to the most isolated areas.

3.2. House Materials

Most of the houses in the studied communities are built of zinc roofs, and wooden walls and floors, and most of them are built on pillars.

3.3. House Bedrooms

The surveys affirm that an average of 5.3 people live under the same roof, where the average bedroom sleeps 2.6 people. The overcrowding index is 2.04 people per room; consequently, it is considered there is “no overcrowding”.

3.4. Windows and Mosquito Nets

74% of the families have no adequate ventilation at home since they do not have wide windows. Additionally, the exposure to tropical disease vectors increases for 43% of the population who do not use mosquito nets around their beds.

3.5. Trash Management

As for the non-organic trash, 40% of the population burn it, 36.7% bury it, and 23% throw it away. 98% of the families use food leftovers to feed their animals. The trash is buried at an average distance of 0.02 miles away from their houses. The maximum distance is 0.18 miles.

3.6. Water Sources for Human Consumption

The main water sources for human consumption are wells (25%) and rain (21.5%). The river continues to be an important source for the population. All of these water sources for human consumption, excepting covered wells and tap water (10%) or drinking water (4.2%), are exposed to any pollutant agent in the environment; consequently, the aerial spraying had a deep impact on people’s health (Figure 4.4).

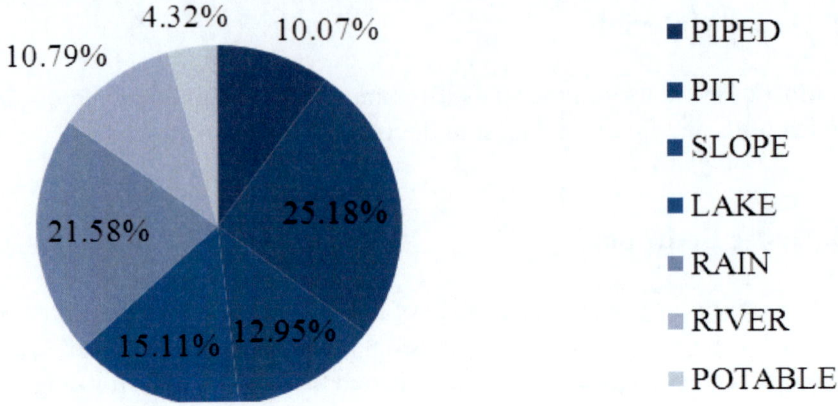

Figure 4.4. Water collection sources.

3.7. Feces Management

38.4% of the population have latrines, whereas 61.4% use the environment. The average distance from the houses to the latrines is 0.03 miles.

3.8. Land Ownership and Use

30.29% of the families from the studied areas have property deeds. Most of them are possessionary (69.7%). The average area of a family farm is 21.3 hectares, although the range of distribution of the properties goes from those who have only a plot to those who have 250 hectares. The families on average take up half property for crops, whereas 2.5% have no farms. For the analysis of the products that are mostly harvested in the area, this 6.9% were excluded. Cacao, pasture, coffee, and short-cycle crops are the main products harvested in the area. These main products are on average assigned around 4 hectares. Cacao, coffee, and banana were mentioned as secondary crops. Coffee is the product that families mostly grow, although cacao and pasture have ousted coffee in terms of land extension. Finally, among the short-cycle crops, families mostly grow corn.

REFERENCES

Instituto Nacional de Estadísticas y Censos (INEC). Censos Nacionales del año 2000 [On line]. 2001. San José, Costa Rica. Available from: URL: www.inec.go.cr/INEC2/censo2000.pdf.

Maldonado, A; Saravia, A; Valladares, C; Álvarez, S; Bonilla, M; Maldonado, P; Supliguicha, V. Estudio de impacto de la actividad petrolera de Perenco, en el bloque 7 de la Región Amazónica Ecuatoriana. *Acción Ecológica*. 2008.

Chapter 5

GENETIC DIAGNOSIS

ABSTRACT

This chapter discusses the genetic diagnoses made regarding the communities affected by the aerial spraying with glyphosate in the province of Sucumbios. As for the genetic analysis, cytogenetic and genetic aspects were evaluated. In connection with the cytogenetic evaluation, the analysis of chromosomes was carried out, starting from chromosome harvesting, culturing, and banding. Regarding genetic analysis, three genes were evaluated: GPX-1, GSTP1 and XRCC1. These genes have detoxification and DNA reparation functions. The evaluation of the presence of genetic polymorphisms was conducted from the molecular technique called polymerase chain reaction – restriction fragment length polymorphism, and the statistical analysis was made by means of the odds ratio technique.

1. BIOLOGICAL SAMPLES

In order to conduct an analysis of the chromosomal aberration and the frequency calculus for the GPX-1, GSTP1 detoxification genes and the XRCC1 repair gene, peripheral blood was extracted through venous tap, using vacutainers with heparin and EDTA (Ethylenediaminetetraacetic acid) as anticoagulants of 92 individuals exposed to the aerial spraying with glyphosate. As control individuals 90 samples of the DNA bank at the Biomedicine Research Institute were analyzed. These individuals had no history of smoking, alcohol, diseases such as cancer, or exposure to carcinogenic agents like pesticides. Each one of these analyzed individuals filled out a biomedical survey and signed their respective informed consent form.

2. Molecular Biology Techniques

2.1. DNA Extraction

The DNA extraction process was conducted starting from the samples of peripheral blood taken from the studied individuals by means of the Purelink™ Genomic DNA Invitrogen™ kit and several steps were standardized at the lab in order to obtain a good pollutant-free DNA concentration and.

The steps to extract DNA are the following:

- A double boiler is prepared at 131° F.
- 200 μL of peripheral blood sample is placed into a 1.5mL sterile Eppendorf tube and 400 μL of saline solution is added to wash the sample and to eliminate as many proteins and enzymes as possible that can inhibit the PCR.
- The sample is mixed on a vortex and is centrifuged for 1 minute to 5.000 g.
- The supernatant is discarded and the sample is subsequently washed until it is transparent.
- 20 μL of Proteinase K is added to the washed blood sample in order to break down proteins; 20μL of RNasa is added to break down the RNA chain. This is mixed in the vortex mixer and incubated at room temperature.
- 200 μL of Genomic Lysis/Binding Buffer is added and is well mixed to obtain a homogenous solution.
- The sample is incubated at 131° F for 10 minutes to remove protein digestion.
- 200 μL of 95% absolute ethanol is added to the cell lysis. It is well mixed in the vortex mixer for 5 seconds in order to obtain a homogenous sample.
- The columns are set up on the collection tubes.

- The 640 µL of cell lysis is added to the membrane column. The column is centrifuged at 10.000 g for 1 minute at room temperature.
- The collection tube is thrown away and the column is placed on a new collection tube.
- 500 µL of Wash Buffer 1 is added to the column previously prepared with ethanol. The column is centrifuged at 10.000 g for 1 minute.
- The collection tube is thrown away and the column is placed on a new collection tube.
- 500 µL of Wash Buffer 2 is added to the column which is centrifuged at maximum speed for 3 minutes at room temperature. The collection tube is discarded.
- The column is placed in a 1.5 mL sterile Eppendorf tube.
- 50 µL of Genomic Elution Buffer is added to the column.
- The sample is incubated at room temperature for 1 minute. The column is centrifuged at maximum speed at room temperature for 1 minute.
- DNA aliquots are made and are kept at -4° F.

2.2. Polymerase Chain Reaction

Polymerase Chain Reaction, also known as PCR, is a molecular biological technique described in 1986 by Kary Mullis (Bartlett & Stirling, 2003). This technique is widely used by researchers around the world due to its high reliability.

The PCR reaction is a cloning method in vitro that is used in the amplification of small DNA quantities. It depends on the polymerase DNA enzyme that resists a temperature of 203° F and has an optimum growth temperature at 161.6° F. The DNA polymerase can only synthesize new DNA for extension of the extreme 3´ end of a preexisting primer. This feature is very commonly used to specifically expand a region between two places determined by the primers. In each cycle the DNA segment is denatured for increase of temperature at 203° F, establishing the primers at an optimum standardized temperature at the lab and incubating at 161.6° F for the polymerase to lengthen the DNA chain. After several cycles, several DNA segment copies are obtained. The reaction takes place on a thermocycler (Sudbery, 2004).

The genomic DNA obtained from the peripheral blood samples taken from individuals exposed to glyphosate is used as a template to enlarge the GPX-1,

GSTP1 detoxification genes and the XRCC1 repair gene, so that the genotype and the presence of the Pro198Leu, Ile105Val, and Arg399Gln polymorphisms can be respectively determined.

2.3. Restriction Fragment Length Polymorphism (RFLP)

The first generation of DNA molecular markers were the restriction fragment length polymorphisms (RFLP). The PCR-RFLP technique is a biallelic genetic marker based on the presence or absence of a sequence; these molecular markers are generally used from the amplification of DNA samples through PCR.

The PCR product is incubated with the specific restriction enzyme and subsequently analyzed in order to observe whether the DNA fragment has been cut (Strachan & Read, 1999). In connection with the likeness between two individuals of the same species, there are individual variations in the sequence of the genetic information. The polymorphisms consist of variations in the DNA sequence.

The restriction enzymes demonstrate the presence or absence of the single nucleotide polymorphisms (SNPs), since any change in the DNA sequence may create or destroy a cut site, leading to a fragment (absence of the cut site) or two fragments (presence of the cut site). The RFLPs are analyzed on the amplicon obtained through the PCR; that is where the name of the PCR-RFLP comes from (Pan, 2005; Riely, 2006).

3. GPX-1 Gene

The experience in the study of polymorphisms in the Ecuadorian population has allowed us to understand how important it is to conduct genetic research in people affected with diseases or exposed to carcinogenic agents. The study of a determined gene basically depends on the said gene and the relationship of the disease or the foreign agent under analysis. For the genetic analysis on the people who live in the communities near the border line with Colombia, the study of detoxification genes (GPX-1, GSTP1) and the DNA repair gene (XRCC1) was determined.

The detoxification genes are interesting when it is about the elimination of free radicals from the body, which, when accumulated, generate high risk of developing diseases such as cancer. The repair genes are essential in the molecular processes of the body and their goal is to correct the bases that have been wrongly attached to the DNA sequence and the poor functioning of this machinery may have an impact on cell problems.

The glutathione peroxidase family (GPX) is made up of six isoenzymes (GPX-1, GPX-2, GPX-3, GPX-4, GPX-5, and GPX-6). The glutathione peroxidase enzymes are so named due to the use of glutathione as an organic chemical to reduce substrates. The GPX-1 antioxidant enzyme is part of the enzymatic antioxidant defense system which prevents oxidative damage to DNA, proteins, and lipids (Ravn-Haren et al., 2006). The characteristics of the GPX-1 gene are detailed in Table 5.1.

Oxidative stress is characterized by the reactive oxygen species (ROS) and the capacity to reduce the activity of antioxidant enzymes which cause metabolic dysfunctions and damage to the biological macromolecules. The accumulation of ROS plays an important role in the pathogenicity of cardiovascular diseases, neurodegenerative disorders, chronic damage to liver and kidneys, as well as different types of cancer (Armstrong et al., 2006).

The enzymes that are in charge of detoxification deactivate compounds and highly dangerous anions for cells. The defenses against reactive oxygen species are made up of glutathione peroxidase (GPX), catalase (CAT), and peroxide dismutase (SOD). GPX-1 is a detoxification enzyme that depends on selenium, protects cells from oxidative damage, reducing hydrogen peroxide and a wide array of organic peroxides. The GPX-1 gene is located in the chromosome region 3p21.

This enzyme is the final regulator of the metabolic pathway that breaks down reactive oxygen species. The GPX-1 protein weighs 21946 Da and is made up of 201 amino acids that are codified through processes of transcription and information translation found in the DNA. It has been reported that the substitution of a nucleotide at codon 198 of the GPX-1 gene results in the transversion of a thymine to a cytosine, leading to the substitution of a leucine (L) to a proline (P) at amino acid level. Moreover, this change is associated with a higher risk of acquiring bladder and lung cancer in Caucasian people due to the reduction of the enzymatic activity production.

Table 5.1. Characteristics of the GPX-1 gene

GPX-1	
Official Symbol	GPX-1
Name	*Glutathione peroxidase 1*
Other Aliases	GSHPX1, MGC14399, MGC88245
Chromosome	3
Localization	3p21.3
NCBI Access Number	NC_000003.11
ID Gene	2878
Sequence Size	1186 bp
Protein Size	201 aa; 21946 Da
Subunit	Homotetramer
Subcellular Localization	Cytoplasm

3.1. Amplification and Genotyping

For the amplification of the 191 bp fragment of the GPX-1 gene, a final PCR volume of 25 μL was obtained. Each sample consisted of 10.85 μL of H_2O Milli-Q, 2.5 μL of buffer 10x (200 mM Tris HCl (pH 8.4), 500 mM KCl), 0.8 μL of $MgCl_2$ 25 mM, 2 μL of dNTPs (dideoxynucleotide triphosphates) 200 mM, 2.5 μL of the primers FW 5´-AAGGTGTTCCTCCCTCGTAGGT-3´ and RV 5´-CTACGCAGGTACAGCCGCCGCT-3´ 0.5 μM, 2.5 units (U) of Taq polimerase and 4 μL of DNA sample at a concentration of 100 ng/μL. After obtaining the PCR reaction, the samples were subsequently placed on the MJ Research PTC® Thermal Cycler for amplification. The first step of initial denaturation lasted 8 minutes at 201.2° F, followed by 36 cycles of 1 minute each at 201.2° F, 1 minute at 140° F, 1 minute at 161.6° F, and a final 9-minute elongation at 161.6° F. The amplification of the fragment was observed through electrophoresis on agarose gel at 2% with ethidium bromide; the DNA bands were observed through the UV transilluminator.

The genotype of the individuals exposed to the aerial spraying with glyphosate was determined by means of the PCR-RFLP technique. The ApaI restriction enzyme was used to identify the Pro198Leu variant. When the restriction enzyme does not recognize the cut site, this is a non-mutated DNA; consequently, a 191 bp band is observed on the agarose gel and it is demonstrated that this is a homozygous individual (P/P). When three bands (191, 117, and 74 bp) are observed, then it is a heterozygous individual (P/L), and when two bands (117 and 74 bp) are observed, it is a mutant homozygous (L/L).

For the restriction enzymes to cut the restriction places, the PCR product was digested during 2 hours at 98.6°F with 5 U of ApaI enzymes. Following the digestion with enzymes, the genotype of each individual was subsequently analyzed through electrophoresis on agarose gel at 5% with ethidium bromide under the UV transilluminator.

3.2. Pro198Leu Polymorphism

Among the individuals exposed to the aerial spraying with glyphosate, it was observed that 35% were normal homozygous with genotype (P/P), 48% were heterozygous with genotype (P/L), and 17% were mutant homozygous with genotype (L/L). The normal allele frequency (P) was 0.59 and the mutant allele frequency (L) was 0.41. Regarding the control individuals, it was observed that 38% were normal homozygous (P/P), 60% were heterozygous (P/L), and 2% were mutant homozygous (P/L). The normal allele frequency was 0.68 and the mutant allele frequency was 0.32 (Table 5.2).

As for the statistical analysis, the results were analyzed by means of two tests; the chi-square (χ^2) and the odds ratio (OR) test. The χ^2 test was applied in order to determine outstanding differences between the presence of the polymorphism and the exposition to glyphosate. It was conducted at a significance level of 0.001 and 2 degrees of freedom. The OR test made it possible to determine the association that exists between the presence of the polymorphism, the exposition to glyphosate, and the risk of developing diseases. The OR analysis was made through a 2 x 2 contingency table.

Table 5.2. Genotyping distribution and allele frequency of the Pro198Leu

Group	Frequency	Individuals	%	Genotyping frequency	Allele frequency
Exposed to glyphosate (n = 92)	P/P	32	35	0.35	0.59
	P/L	44	48	0.48	
	L/L	16	17	0.17	0.41
Control (n = 90)	P/P	34	38	0.38	0.68
	P/L	54	60	0.6	
	L/L	2	2	0.02	0.32

Regarding the Pro198Leu polymorphism of the GPX-1 gene, an OR of 0.87 for the genotype P/L (OR = 0.87; 95% IC 0.5-1.6; $p = 0.77^{NS}$), an OR of 8.5 for the genotype L/L (OR = 8.5; 95% IC 1.8-39.9; $p < 0.05^{*}$), and an OR of 1.14 for the combination of the genotypes P/L + L/L (OR = 1.14; 95% IC 0.6-2.1; $p = 0.79^{NS}$) were obtained (Table 5.3). The results allow us to determine that the individuals with the presence of the genotype L/L have a significant association between the presence of the Pro198Leu polymorphism, the alteration of the levels of glutathione peroxidase enzyme, and the possible acquisition of problems in the process of peroxide detoxification at cell level.

4. GSTP1 GENE

The glutathione S-transferase pi 1 gene (GSTP1) belongs to the cytosolic superfamily GST (Cowell et al., 1988). This gene is located in the chromosome 11 in the 11q2 region and is made up of a DNA sequence of 3066 bp (Table 5.4).

The GSTP1 gene codifies for a phase II enzyme of 210 amino acids which catalyzes the tripeptide glutathione conjugation with a wide array of electrophilic compounds, including genotoxics, carcinogens, and chemotherapeutic agents (Lo et al., 2008). This enzyme detoxificates free radicals, especially tobacco subtracts, and interacts with an array of electrophilic xenobiotics, including subtracts that go from environmental toxins, carcinogens to drugs used in cancer treatments (Ketterer, 2001; Hayes et al., 2005; Zendehdel et al., 2009). The GSTP1 gene is also responsible for the detoxification of reactive metabolites, accelerating the excretion range (Rybicki et al., 2006).

The glutathione S-transferase pi 1 enzyme not only functions as a phase II drug metabolizer, but also as an apoptosis regulator (Moyer et al., 2008), as a kinase pathway regulator MAPKs as a result of the enzymatic activity (Wu et al., 2006; Zhao et al., 2006), as a regulator of cell signals through the connection to proteins, such as c-Jun NH2- terminal kinase, ASK1, and TRAF2, regulating the *downstream* signal of the genes (Wu et al., 2006). When the activity of the GSTP1 gene decreases, the cells become more vulnerable to mutations due to the oxidative stress (Meiers et al., 2007).

A (A-to-G) transition mutation in the nucleotide 313, in the exon 5 of the GSTP1 gene has been reported. This DNA mutation results in the translation of the valine amino acid (V), instead of isoleucine (I), in the codon 105 of the active site of the enzyme. This mutation is associated with the reduction of the enzymatic activity, altering its thermostability. The presence of the allele V has been related to an increase of the risk of acquiring bladder, testicular, and prostate cancer (Zendehdel et al., 2009).

Table 5.3. Odds Ratio statistical analysis in the GPX-1, GSTP1, and XRCC1 genes

Genes	Genotype	OR	95% IC	*p*
GPX-1 Pro198Leu	P/P	1.0 (Reference)		
	P/L	0.87	0.5 – 1.6	0.77^{NS}
	L/L	8.5	1.8 – 39.9	< 0.05*
	P/L + L/L	1.14	0.6 – 2.1	0.79^{NS}
GSTP1 Ile105Val	I/I	1.0 (Reference)		
	I/V	1.95	1.0 – 3.8	0.07^{NS}
	V/V	4.88	2.0 – 11.8	< 0.001*
	I/V + V/V	2.6	1.4 – 4.8	< 0.05*
XRCC1 Arg399Gln	R/R	1.0 (Reference)		
	R/Q	12.2	0.7 – 219.8	0.4^{NS}
	Q/Q	0.03	0.003 – 0.2	< 0.001*
	R/Q + Q/Q	0.2	0.02 – 1.4	0.1^{NS}

Table 5.4. Characteristics of the GSTP1 gene

GSTP1	
Official Symbol	GSTP1
Name	*Glutathione S-transferase pi 1*
Other Aliases	DFN7, FAEES3, GST3, GSTP, PI
Chromosome	11
Localization	11q2
NCBI Access Number	NC_000011.9
ID Gene	2950
Sequence Size	3066 bp
Protein Size	210 aa

4.1. Amplification and Genotyping

For the amplification of the 177 bp fragment of the GSTP1 gene, a final PCR volume of 50 μL was obtained. Each sample consisted of 37 μL of H_2O

Milli-Q, 5 μL of buffer 10x, 1.5 μL of $MgCl_2$ 25 mM, 1 μL of dNTPs 200 mM, 2.5 μL of the primers FW 5´-ACCCCAGGGCTCTATGGGAA-3´ and RV 5´-TGAGGGCACAAGAAGCCCCT-3´ at 0.5 μM, 2.5 U of Taq polymerase, and 4 μL of ADN. The conditions of the amplification consisted of an initial denaturation for 5 minutes at 203° F, followed by 35 cycles of 45 seconds at 201.2° F, 1 min at 143.6° F, 45 seconds at 161.6° F, and a final elongation for 5 minutes at 161.6° F. The amplicon migrated on an agarose gel at 2% with ethidium bromide in an electrophoresis camera, and the bands were observed through a UV transilluminator.

The HaeIII restriction enzyme was used to determine the Ile105Val polymorphism of the individuals in the study. When the enzyme recognizes the cut site, a 177 bp band is observed and it is proven that this is a homozygous individual (I/I). When three bands (177, 92, and 85 bp) are observed, then it is a heterozygous individual (I/V), and when two bands (92 and 85 bp) are observed, this is a mutant homozygous (V/V). The HaeIII restriction enzyme led the PCR product for 2 hours at 98.6°F with 5 U. The bands were observed on agarose gel at 5% with ethidium bromide under the UV light.

4.2. The Ile105Val Polymorphism

Regarding the individuals exposed to the aerial spraying with glyphosate, it was observed that 32% were normal homozygous (I/I), 40% were heterozygous (I/V), and 28% were mutant homozygous (V/V). The normal allele frequency (I) was 0.52 and the mutant allele frequency (V) was 0.48. As for the control individuals, it was observed that 54% were normal homozygous (I/I) and 10% were mutant homozygous (V/V). The normal allele frequency was 0.72 and the mutant allele frequency was 0.28 (Table 5.5).

Concerning the statistical analysis of the Ile105Val polymorphism of the GSTP1 gene, an OR of 1.95 for the genotype I/V (OR=1.95; 95% IC 1.0-3.8; $p = 0.07^{NS}$), an OR of 4.88 for the genotype V/V (OR = 4.88; 95% IC 2.0-11.8; $p < 0.001$*), and an OR of 2.6 for the combination of genotypes I/V + V/V (OR=2.6; 95% IC 1.4-4.8; $p < 0.05$*) were obtained (Table 5.5). These results show us that the individuals with genotype V/V have a greater risk of acquiring problems in the detoxification processes of metabolites, as well as in the cell signaling process for apoptosis and activation of kinase genes.

5. XRCC1 GENE

The XRCC1 gene is one of more than twenty genes that participate in the base excision repair; it is located in the chromosome 19 in the 19q13.2 region, it codifies a scaffold protein of 633 amino acids, and it has a size of 69526 Da which plays a role in the DNA rupture repairs, one of the most common injuries in genetic material (Table 5.6) (Caldecott et al., 1995).

Human cancer may be developed due to the damage in the DNA caused by exposure to UV rays, ionizing radiation, and environmental chemical agents. In order to protect the integrity of a genome, human beings have developed a complex DNA repair system. Among the functions of the DNA, the repair by excision is the main defensive mechanism against the damage as a result of the cell metabolism, including reactive oxygen species, methylation, deamination, and hydroxylation. Consequently, the base excision repair is a universal event in the cells and is relevant in the prevention of mutagenesis (Zhibin et al., 2005). The XRCC1 gene interacts with a group of DNA repair proteins, including poly (ADP-ribose) polymerase, DNA Ligase 3, and DNA polymerase-h (Caldecott et al., 1995; Dianov et al., 1999; Thompson et al., 2000).

Since the Arg399Gln polymorphism is located in the BRCT-1 domain of the XRCC1 gene inside the poly (ADP-ribose) region, it has been widely researched both in its function and its association with the risk of getting cancer. Therefore, the study of frequencies of the Arg399Gln polymorphism of the XRCC1 gene in the population exposed to the aerial spraying with glyphosate is extremely important.

Table 5.5. Genotyping distribution and allele frequency of Ile105Val

Group	Frequency	Individuals	%	Genotyping frequency	Allele frequency
Exposed to glyphosate (n = 92)	I/I	29	32	0.32	0.52
	I/V	37	40	0.40	
	V/V	26	28	0.28	0.48
Control (n = 90)	I/I	49	54	0.54	0.72
	I/V	32	36	0.36	
	V/V	9	10	0.10	0.28

Table 5.6. Characteristics of the XRCC1 gene

XRCC1	
Official symbol	XRCC1
Name	*X-ray repair cross complementing protein*
Other aliases	RCC
Chromosome	19
Localization	19q13.2
NCBI Access Number	NC_000019.9
ID Gene	7515
Protein Size	633 aa; 69526 Da
Subunit	Homodimer
Subcellular Localization	Nucleus

The Arg399Gln polymorphism consists of the substitution of the G nitrogenous base for A in the codon 399, exon 10 of the XRCC1 gene; being the change of arginine (R) for glutamate (Q) at the amino acid level. The presence of this variant has been associated with the reduction of the DNA capacity to repair mutations for the presence of DNA adducts (Lunn et al., 1999; Matullo et al., 2001).

5.1. Amplification and Genotyping

The amplification of the 242 bp fragment of the XRCC1 gene was made from a final PCR volume of 50 μL. Each sample was made up of 34 μL of H_2O Milli-Q, 5 μL of buffer 10x, 1.5 μL of $MgCl_2$ 25 mM, 1 μL of dNTPs 200 mM, 2 μL of the primers FW 5´-CCCCAAGTACAGCCAGGTC-3´ and RV 5´-TGCCCCGCTCCTCTCAGTAG-3´ at 0.4 μM, 2.5 U of Taq polymerase, and 4 μL of DNA. The conditions for the amplification of the XRCC1 gene were 203° F for 5 minutes, 35 cycles of 201.2° F for 15 seconds, 138.2° F for 1 minute, 161.6° F for 30 seconds, and 161.6° F for 5 minutes. The amplified bands were observed under the UV light.

To be able to determine the presence of the Arg399Gln polymorphism, the MspI restriction enzyme was used. When the enzyme does not recognize the cut site, a 242 bp fragment is observed; this, then, is a homozygous individual (R/R). When three fragments (242, 148, and 94 bp) are observed, this is a heterozygous individual (R/Q), and when two fragments (148 and 94 bp) are observed, this is a mutant homozygous individual (Q/Q). Likewise, the amplicons are digested with 5U of the MspI enzyme for 2 hours at 98.6° F. Also, the bands were observed under the UV light.

5.2. The Arg399Gln Polymorphism

Regarding the individuals exposed to glyphosate, it was observed that 7% were normal homozygous individuals (R/R), 79% were heterozygous individuals (R/Q), and 14% were mutant homozygous individuals (Q/Q). The normal allele frequency (R) was 0.46 and the mutant allele frequency (Q) was 0.54. As for the control individuals, it was observed that 1% were normal homozygous individuals (R/R), 1% were heterozygous individuals (R/Q), and 98% were mutant homozygous individuals (Q/Q). The normal allele frequency was 0.02 and the mutant allele frequency was 0.98 (Table 5.7).

Concerning the Arg399Gln polymorphism of the XRCC1 gene, an OR of 12.2 was obtained for the genotype R/Q (OR = 12.2; 95% IC 0.7-219.8; $p = 0{,}4^{NS}$), an OR of 0.03 for the genotype Q/Q (OR = 0.03; 95% IC 0.003-0.2; $p < 0.001$*), and an OR of 0.2 for the combination of the genotypes R/Q + Q/Q (OR = 0.2; 95% IC 0.02-1.4; $p = 0.1^{NS}$) (Table 5.7). The data that was obtained proves that there is no association or risk between the Arg399Gln polymorphism and damage in the DNA in the studied individuals, demonstrating as a result a stable DNA repair system.

Table 5.7. Genotyping distribution and allele frequency of Arg399Gln

Group	Frequency	Individuals	%	Genotyping frequency	Allele frequency
Exposed to glyphosate (n = 92)	R/R	6	7	0.07	0.46
	R/Q	73	79	0.79	
	Q/Q	13	14	0.14	0.54
Control (n = 90)	R/R	1	1	0.01	0.02
	R/Q	1	1	0.01	
	Q/Q	88	98	0.98	0.98

6. Cytogenetics

The exposure of the eukaryotic cells to several chemical and physical agents may prompt damage or alterations at chromosome level (Au, 1991;

Leonard et al., 1985). A cytogenetic analysis makes it possible to learn about the numerous and structural alterations found in the chromosomes of those individuals exposed to the aerial spraying with glyphosate.

Obtaining karyotype from each one of the individuals consists of stimulating cell growth through specific culture means, culturing the cells and holding them in metaphase, distributing the metaphase on plaques, dyeing, and banding the chromosomes in order to subsequently analyze them in a microscope with a specialized computer program.

6.1. Peripheral Blood Culturing

The peripheral blood culturing consists of placing the cells in a favorable environment so as to stimulate their division; in this way, adequate chromosomes are obtained for further analysis.

The steps that have been taken for culturing are the following:

- The laminar flow cabinet is cleaned up with bleach and alcohol.
- The aliquot of the RPMI 1640 culturing medium is defrosted through a double boiler at 98.6° F.
- The patient code and the proceeding date are written on a new conical tube.
- The conical tube and the flask containing the RPMI 1640 medium that was defrosted are placed in the laminar flow cabinet and the UV light is turned on for 10 minutes before culturing.
- A lab coat, a mask, and gloves disinfected with alcohol are worn to carry out the culturing process.
- 5 mL of RPMI 1640 medium is placed into the conical tube and 10 drops of the sample (around 1 mL) are added.
- The conical tube containing the culture fluid and all the reactives used with parafilm are sealed.
- The sample is incubated on the heater for 48 hours at 98.6° F in order to observe genotoxicity.
- The cabinet is cleaned up and, if there is left-over, it is stored in the refrigerator at 39.2° F.

6.2. Peripheral Blood Harvesting

The harvesting of the cells that were previously cultured is divided by two parts: the first part consists of adding colchicine to the sample, in order to hold the cell division in metaphase; this way, the chromosomes can be better observed. The second part consists of breaking the cell membranes to release the chromosomes in metaphase, so that the staining and the subsequent analysis can come afterwards.

The steps taken to harvest the blood samples are the following:

- 200 μL of Colcemid® is added and the sample is placed on the heater at 98.6° F for 1 hour.
- In the meantime, the hypotonic shock and Carnoy's solution are prepared.
- The amount of hypotonic shock that is going to be used is heated on double boiler at 98.6° F.
- Once the colchicine time is up, the sample is centrifuged at 2500 rpm for 10 minutes.
- The supernatant is discarded up to the line of the conical tube (1mL) in a flask containing diluted bleach, and the button is resuspended.
- The entire sample in the pipette is absorbed and 7mL of hypotonic shock at 98.6° F is placed into the empty tube.
- The fluid in the pipette is slowly released into the tube and it is incubated in blood on double boiler at 98.6° F for 25 minutes.
- Following the time of the double boiler, the sample is centrifuged at 2500 rpm for 10 minutes.
- The supernatant is discarded up to the line of the conical tube and the button is resuspended.
- The entire sample in the pipette is absorbed and 4 mL of solution is placed into the empty tube.
- The fluid in the pipette is quickly released; it is resuspended 2 or 3 times with the pipette.
- The sample is settled at room temperature for 15 minutes.
- Once the time is up, the sample is centrifuged at 2500 rpm for 10 minutes and the supernatant is discarded up to the line of the conical tube and the button is resuspended. The sample in the pipette is absorbed and 4 mL of solution is placed into an empty tube, the solution in the pipette is quickly released; it is 2 or 3 times resuspended with the pipette.

- The sample is centrifuged at 2500 rpm for 10 minutes, the supernatant is discarded up to the line of the conical tube and the button is resuspended.
- The wash with the solution is repeated until a clean pellet is obtained.
- The sample is settled in the refrigerator for 1 hour before extending the plaques.

6.3. Plaque Extension

The process of plaque extension consists of permitting a good distribution of the metaphase throughout the plaque.

The steps involved in plaque extension are the following:

- The sample is taken out of the refrigerator and it is centrifuged at 2500 rpm for 10 minutes.
- After immersing the plaque in solution at 60% of acetic acid, the plaque is marked, checking the number of the last extended plaque, in accordance with the respective registration.
- After the centrifugation time, the supernatant is discarded until 3mL of the sample is remaining and the button is resuspended. In case of obtaining more than 1mL of pellet, it is gently resuspended with the Pasteur pipette, raising the cells from the surface.
- The sample is dropped on the plaque from a 12-inch height, 4 drops of the resuspended simple (abundant pellet), 6 drops (in case of normal pellet), and 8 drops (in case of scarce pellet).
- As for direct plaque staining, the sample is quickly placed on fire. In case of the plaques for GTG-banding, the sample is placed on a tray and taken to the heater at 140° F for 2 days.
- The plaques are dyed.

6.4. Simple Plaque Staining

Simple plaque staining consists of coloring the chromosomes with Giemsa, in order to observe the metaphases to be analyzed.

The steps for staining are the following:

- A Giemsa solution is prepared in a Coplin glass (5mL of mother solution and 45 mL of water).
- The plaques that are already extended are placed in the Coplin glass for 8 to 10 minutes.
- The plaque is taken through a water stream in order to remove the colorant excess, to keep it dry, and to observe with a microscope.

6.5. GTG-Banding

The chromosome GTG-banding makes it possible to create a banding pattern in the short and long arms in order to facilitate the identification of the homologous chromosome pair; this way, it makes it possible to make the karyotype of the healthy individuals and those who were exposed to the aerial spraying with glyphosate.

The GTG-banding process of the plaques is the following:

- The plaques are aged at 140° F for 48 hours.
- The trypsin is defrosted and is placed on double boiler at 98.6° F.
- 45mL of water in the Coplin glass is heated up until 98.6° F is reached. Once that temperature is reached, 5mL of trypsin is added.
- The plaque is placed in the trypsin solution at 98.6° F for 30 minutes.
- The plaque is washed with NaC1 at 0.95%, and subsequently it is washed with water.
- The plaques are stained according to the simple staining protocol with Giemsa.
- The plaques are dried at room temperature and are observed with the microscope.

6.6. Karyotype Analysis

The metaphases of each individual were analyzed with a 100X lens in an Olympus BX51 microscope. The karyotype of the people exposed to the aerial spraying was made with the aid of the CytoVision® System software (Figure 5.7). The results show that 0% of the individuals had chromosome alterations and less than 3% of chromosome fragility was observed; in other words, all of the individuals were within the normal established range. The results are detailed in Table 5.8.

REFERENCES

Armstrong, V; Atanasova, S; Oellerich, M; Schlumbohm, C; Von Ahsen, N; Wieland, E. Marmoset glutathione peroxidases: cDNA sequences, molecular evolution, and gene expression. *J Med Primatol.*, 2006, 35, 155-164.

Au W. Monitoring human populations for effects of radiation and chemicals exposures using cytogenetic techniques. *Occ Med.*, 1991, 6(4).

Bartlet J; Stirling D. A short history of the polymerase chain reaction. *Meth Mol Biol.*, 2003, 226(3-6).

Caldecott, K; Stanker, L; Thompson, L; Tucker, J. Characterization of the XRCC1-DNA ligase III complex in vitro and its absence from mutant hamster cells. *Nucleic Acids Res.*, 1995, 23, 4836-4843.

Cowell, I; Dixon, K; Pemble, S; Ketter, B; Taylor, J. 1988. The structure of the human glutathione S transferase pi gene. *Biochem J.*, 1988, 25, 79-83.

Dianov, G; Prasad, R; Wilson, S; Bohr, V. Role of DNA polymerase h in the excision step of long patch mammalian base excision repair. *J. BiolChem.*, 1999, 27, 13741-13743.

Hayes, J; Flanagan, J; Jowsey, I. Glutathione transferases. *Annu Rev Pharmacol Toxicol.*, 2001, 45, 51-88.

Ketterer, B. A bird's eye view of the glutathione transferase field. *Chem Biol Interact.*, 2001, 13, 27-42.

Leonard, A. Population monitoring for genetic damage induced by environmental physical and chemical agents. *Environ Monit Assess.*, 1985, 5, 69-284.

Lo, H; Stephenson, L; Cao, X; Milas, M; Pollock, R; Ali-Osman, F. Identification and functional characterization of the human glutathione S transferase P1 gene as a novel transcriptional target of the p53 tumor suppressor gene. *Mol Cancer Res.*, 2008 6(5), 843-850.

Lunn, R; Langlois, R; Hsieh, L; Thompson, C; Bell, D. XRCC1 polymorphisms: effects on aflatoxin B1-DNA adducts and glycophorin A variant frequency. *Cancer Res.*, 1999, 59, 2557-2561.

Matullo, G; Guarrera, S; Carturan, S. DNA repair gene polymorphisms, bulky DNA adducts in white blood cells and bladder cancer in a case-control study. *Int J Cancer.*, 2001, 92, 562-567.

Meiers, I; Shanks, J; Bostwick, D. Glutathione S-transferase pi (GSTP1) hypermethylation in prostate cancer: review. *Pathology.*, 2007, 39, 299-304.

Moyer, A; Salavaggione, O; Wu, Y; Moon, I; Eckloff, B; Hildebrandt, M; Schaid, D; Wieben, E; Weinshilboum, R. Glutathione S-transferase P1: Gene sequence variation and functional genomics studies. *Cancer Res.*, 2008, 68, 12, 4791-4801.

Pan, Q; Pao, W; Ladanyi, M. Rapid polymerase chain reaction-based, detection of epidermal growth factor receptor gene mutations in lung adenocarcinomas. *J Mol Diagn.*, 2005, 7(3), 396-403.

Ravn-Haren, G; Olsen, A; Tjonneland, A; Dragsted, L; Nexo, B; Wallin, H; Overvad, K; Raaschou-Nielsen, O; Vogel, U. Associations between, GPX1 Pro198Leu polymorphism, erythrocyte GPX activity, alcohol consumption and breast cancer risk in a prospective cohort study. *Carcinogenesis.*, 2006, 27(4), 820-825.

Riely, G; Pao, W; Pham, D; Li, A; Rizvi, N; Venkatraman, E; Zakowski, M; Kris, M; Ladanyi, M; Miller, V. Clinical course of patients, with non-small cell lung cancer and epidermal growth factor receptor exon 19 and exon 21 mutations treated, with Gefitinib or Erlotinib. *Clin Cancer Res.*, 2006, 12(3), 839-844.

Rybicki, B; Neslund-Dudas, C; Nock, N; Schultz, L; Eklund, L; Rosbolt, J; Bock, CH; Monaghan, K. Prostate cancer risk from occupational exposure to polycyclic aromatic hydrocarbons, interacting with the GSTP1 Ile105Val polymorphism. *Cancer Detect Prev.*, 2006, 30(5), 412-422.

Strachan, T; Read, A. Human Molecular Genetics 2. Second Edition. BIOS Scientific Publishers Ltd. Oxford, UK. 1999.

Sudbery, P. Genética Molecular Humana. Segunda Edición. Pearson Prentice Hall. 2004.

Thompson, L & West, M. XRCC1 keeps DNA from getting stranded. *Mutat Res.*, 2000, 459, 1-18.

Wu, Y; Fan, Y; Xue, B. Human glutathione S-transferase P1-1 interacts with TRAF2 and regulates TRAF2-ASK1 signals. *Oncogene.*, 2006, 25, 5787-5800.

Zendehdel, K; Bahmanyar, S; McCarthy, S; Nyren, O; Andersson, B; Ye, W. 2009. Genetic polymorphisms of glutathione S-transferase genes GSTP1, GSTM1, and GSTT1 and risk of esophageal and gastric cardia cancers. *Cancer Causes Control.*, 2009, 10552-10557.

Zendehdel, K; Bahmanyar, S; McCarthy, S; Nyren, O; Andersson, B; Ye, W. Genetic polymorphisms of glutathione S-transferase genes GSTP1, GSTM1, and GSTT1 and risk of esophageal and gastric cardia cancers. *Cancer Causes Control.*, 2009, 10552-10557.

Zhao, X; Fan,Y; Shen, J; Wu, Y; Yin, Z. Human glutathione S-transferase P1 suppresses MEKK1-mediated apoptosis by regulating MEKK1 kinase activity in HEK293 cells. *Mol Cells.*, 2006, 21, 395-400.

Zhibin, H. XRCC1 polymorphisms and cancer risk: A meta-analysis of 38 case-control studies. *Cancer Epidemiol Biomark Prev.*, 2005, 14(7), 1810-1818

Chapter 6

HEALTH DIAGNOSIS

ABSTRACT

This chapter discusses a wide array of information related to the medical examination of the people who live in the communities that were exposed to glyphosate in the Ecuadorian Northern border line. Among the most relevant issues are the nutritional status assessment; the health assessment; the mother-child health status; the relevant personal records, such as intoxication, infectious diseases; habits; impact of the aerial spraying on health; and the presence of symptoms in immunological, visual, digestive, epithelial, hearing, respiratory, cardiovascular, locomotor, blood, endocrine, and central nervous systems. The information that was gathered in this chapter makes it possible to discuss the presence of problems in these people's life quality, which worsens when exposed to genotoxic agents.

1. NUTRITIONAL ASSESSMENT

In order to make a nutritional assessment of the population, three groups were made up for those who are under 18 years old: 0 to 5 years old, 6 to 11 years old, and 12 to 17 years old. As for adults, four groups were created: 15 to 34 years old, 35 to 44 years old, and 45 to 64 years old. The creation of these groups allows for making a more efficient analysis of the nutritional assessment of the studied communities.

1.1. Global Nutritional Status

The global nutritional level was measured by means of the "weight-age" index, which is a good indicator to determine weight insufficiency. It plays an important role in children under 1 year old. This indicator is used to see the normal weight increase in children and is useful to identify the risks of weight loss (Table 6.1).

According to research done on a population made up of 262 children and teens (131 boys and 131 girls), we found an (SM+MM) index of 7.2%, with a risk of malnutrition of 25.2%, and an overweight index (OW+OB) of 8.8%.

This data is better than that found in research done in the bordering area (Maldonado et al., 2006), on which the nutritional status of the school population in communities located less than 2 miles away from the border was studied, where there was malnutrition in 10.3%, a risk of malnutrition of 36.3%, and an overweight index of 4.6%, but that study did not include children younger than 6 years old.

If only the population from 6 to 17 years old that was reviewed in the current research is compared, after two years without the aerial sprayings in connection with the *Plan Colombia*, the nutritional condition has considerably improved, since there is only a 3% of malnutrition, the risk of mild malnutrition (mM) decreases to 23.2%, whereas the normal nutritional status reaches up to 67.1% among the school population, which did not even hit 50% three years ago.

If the malnutrition status is analyzed considering age and sex, it is found that there is more (SM+MM) malnutrition in children under 6 years old, reaching up to 14% in boys and 13.6% in girls. The normal weight does not reach half of the population in boys and girls from this age group, where there was some 48% of normal weight in boys and 40.9% in girls. It is interesting to note that while there is an increasing tendency of mild malnutrition in boys, the tendency decreases in girls, as they grow older. The data and the tendencies that belong to the global nutrition both in boys and in girls from 0 to 5 years old, 5 to 11 years old, and 12 to 17 years old can be observed in Figures 6.4 and 6.5.

The global malnutrition in the country in children under 5 years old, with a decreasing tendency, is a little larger than the numbers obtained from the affected communities (Table 6.2).

Table 6.1. Weight/age clinical category

Weight/age category	Relationship to the median	Meaning
SM: Severe malnutrition	Same and less than -3SD[a]	Dangerous malformation
MM: Moderate malnutrition	From -2SD to -2.99SD	MM, together with the SM, defines the malnutrition level of the population
mM: Mild malnutrition	From -1SD to -1.99SD	Indicates risk of malnutrition
NW: Normal weight	More than -1SD to less than +1SD	Normal weight in relation to age
OW: Overweight	From +1SD to +1.99 SD	Risk of obesity
OB: Obesity	Same and more than +2SD	Obesity

[a]SD: Standard deviation with respect to the median. It is positive (+) when it is larger or negative (-) when the value does not reach the median.

Table 6.2. Distribution of global malnutrition by year, area, region, and sex

Year	Country	Area		Region			Sex	
		Urban	Rural	Highlands	Coast	Amazon Basin	Male	Female
1999[a]	17.2	13.5	21.8	18.5	15.9	-	16.0	18.3
2000[a]	16.4	13.3	20.7	18.0	15.1	17.4	15.2	17.6
2004[a]	14.6	12.4	18.4	14.2	15.0	14.1	16.2	13.0

[a] For children under 5 years old.

1.2. Chronic Nutritional Status (Growth Delay)

This status is measured by means of the "stature-age" indicator. It determines bone growth and it is the best indicator to measure growth delay.

Additionally, it measures adequate diet and nutritional status in the long term. Its data is irreversible and chronicity is indicated (Table 6.3).

For this research the reference data was taken from the list created by the World Health Organization (WHO) in Geneva in 1979 (WHO, 1979). Using this list, all of the 242 people under 18 years old were checked and the following chronic nutrition level was found: a short stature (SS) of 29% (chronic malnutrition indicator), a risk of malnutrition (mildly tall stature) in 40%, a normal stature (NS) in almost a quarter of the population (24%), and tall stature in 7% of the population.

On the research done in 2006 by Maldonado et al., in the bordering communities with the school population (without children under 6 years old), the following distribution of nutritional status was found: a chronic malnutrition of 29% and a risk of 35% in those schools that were located less than 2 miles away from the bordering line. If the population of children under 6 years old is omitted and we compare the data obtained in this research to those obtained in 2006, we find that there are no significant variations, which is a credit for the chronic nature of the indicator not drastically varying after two years' absence of the aerial spraying. The data of this study reflect 28% chronic malnutrition in the school children aged from 6 to 17 years old and a risk of malnutrition of 40%.

As for chronic malnutrition in Ecuador in children under 5 years old, the official data provide the upcoming distribution according to the year:

Table 6.3. Stature/age clinical category

Stature/age category	Relationship to the median	Meaning
SS: Short stature	Same and less than -2 SD[a]	Indicates chronic malnutrition
MSS: Mild short stature	From -1 SD to -1.99 SD	Indicates risk of chronic malnutrition
NS: Normal stature	From -0.99 SD to +0.99 SD	Normal stature
MTS: Mild tall stature	From +1 SD to + 1.99 SD	Slightly over than the median
TS: Tall stature	Same and more than +2 SD	Very tall for the median

[a]SD: Standard deviation with respect to the median. It is positive (+) when it is larger or negative (-) when the value does not reach the median.

Table 6.4. Distribution of chronic malnutrition by year, area, region, and sex

Year	Country	Area		Region			Sex	
		Urban	Rural	Highlands	Coast	Amazon Basin	Male	Female
1999	22.9	14.5	33.6	30.6	15.7	-	22.4	23.4
2000	22.7	15.6	32.8	29.6	17.1	27.2	24.9	20.3
2004	17.4	12.0	26.6	23.0	12.3	21.3	19.5	15.2

There is more malnutrition in children from 0 to 5 years old and teenagers who are growing and developing physically in a definite and accelerated manner. While in the first group there is malnutrition and a risk of 72%, in the group from 6 to 11 years old there is malnutrition and a risk of 62% (10 percentage points less, which is very outstanding) and in the group from 12 to 17 years old there is malnutrition and a risk of 77%, more than three-quarters of the youth population.

If the chronic nutrition levels are compared by sex, it can be observed that there is more malnutrition in girls. After considering the short stature and the mild short stature indicators, boys have a deficiency of 66% and girls 72%.

1.3. Acute Nutritional Status

The weight-stature indicator is known as a good indicator for recent weight loss and gain and shows the recent nutritional status. This is the favorite indicator in emergency situations in order to assess the nutritional loss that this status generates. The low indicators indicate that a recent severe event of weight loss has taken place due to a severe illness or starvation.

Chevassus-Agnes determines that provided there is no important loss of food, it is normal in the population when there is a weight loss (SM+MM) of some 5% even in poor countries. If the rate is larger than 5%, we would be talking about an alarming situation; ranges from 10-14% would indicate such a serious status, and when the rate is larger than 15%, that would be a critical status (Chevassus-Agnes, 1999) (Table 6.5).

Table 6.5. Weight/stature clinical category

Weight/ age category	Relationship to the median	Meaning
SM: Severe malnutrition	Same and less than -3SD[a]	Dangerous malnutrition
MM: Moderate malnutrition	From -2SD to -2.99SD	MM, together with SM, defines malnutrition in the population
mM: Mild malnutrition	From -1SD to -1.99SD	Risk of malnutrition
NW: Normal weight	More than -1SD to less than +1SD	Normal weight for stature although it does not discriminate short children
OW: Overweight	From +1SD to +1.99 SD	Risk of obesity
OB: Obesity	Same and more than +2SD	Obesity

[a]SD: Standard deviation with respect to the median. It is positive (+) when it is larger or negative (-) when the value does not reach the median.

The acute nutritional status of the children and youth population is below the 5% previously mentioned; therefore, the numbers show there is currently no danger.

In order to compare these numbers to those from the research done in 2006, the group from 0 to 5 years old has to be omitted, since the aforementioned research was only based on the school population that lived less than 2 miles away from the bordering line.

The comparison shows a similar pattern although the malnutrition level differs in the percentage. In 2006 there was an acute malnutrition of 1.87% and a mild malnutrition of 7.17%.

The data obtained in 2009 are slightly better for malnutrition, since it goes from an acute malnutrition of 1% to a mild malnutrition of 5.8%. Nevertheless, the overweight and obesity rates are considerably higher in 2009, where obesity increased from 5.92% to 16%, which does not necessarily mean better nutrition, but a significant nutritional imbalance.

This tendency to obesity will also be evident in the adult population by means of measuring the Body Mass Index (BMI).

1.4. Body Mass Index (BMI)

The WHO has determined that the BMI is a good indicator to establish the nutritional status and the nutritional quality in a population or community. The WHO proposes an assessment table which is useful to classify the nutritional status in a population. The BMI is obtained from the division of a person's weight in kilograms by the square of their height in meters (Table 6.6).

The results in the following figures show that in the bordering communities, after two years since the last aerial spraying, there is no malnutrition in the population older than 18 years old, showing a clear tendency to obesity. This tendency is greater in women, 64.1% of whom have some degree of obesity (29.7%) or are overweight, in contrast with 7.8% in men.

The statistics provided by the WHO in 2005 detail that obesity in women in Ecuador was 16.7% and in men 6.7%. In our research it is observed that these statistics are widely exceeded by women and slightly exceeded by men (WHO, 2005). Additionally, it can be highlighted that if our research is compared to that conducted by Maldonado et al., the BMI is very similar to that of the communities that are not affected by the oil community (Maldonado et al., 2008). Regarding age, it is observed that obesity takes place mainly in adult people, so that the percentages show an increasing tendency in this category in both men and women, starting from 35 years old.

Table 6.6. Classification of the Body Mass Index

Classification	Meaning
< 16	Entry criteria (Severe malnutrition)
16 a 16.99	Infra-weight (Moderate malnutrition)
17 a 17.99	Underweight (Mild malnutrition)
18 a 24.99	Normal weight (Healthy)
25 a 29.99	Overweight (Obesity degree I)
30 a 34.99	Chronically overweight (Obesity degree II)
35 a 39.99	Premorbid obesity (Obesity degree III)
40 a 49.99	Morbid obesity (Obesity degree IV)
> 45	Hypermorbid obesity (Obesity degree V)

2. Health Assessment

2.1. Health Perception

The health perceptions in the family surveys of heads of household, two years after the last aerial spraying, indicates that 7% of them consider they have very good or good health, some 57.2% regular, and some 35.7% bad or very bad. Their health perception of the rest of the family is slightly better with some 14.5% who consider their health is very good or good, some 64.9% regular, and some 20.4% bad or very bad.

If these results on health perception are compared to those of other studies conducted in the same Amazon area, it is found that the health perception is better than those perceived in the Texaco and Perenco oil areas (Beristain et al., 2009), although it is lower in those communities that have no oil activity or are not located in the bordering line (Maldonado et al., 2008) (Figure 6.1).

2.2. Family Mortality

The family surveys made it possible to find out that 22.22% of the families had suffered from the death of a relative while living on the farms where they currently live. Out of that total, after reviewing the causes of death, it was observed that some 55.8% of the causes of death are unknown or undiagnosed, in contrast with some 32.35% of cancer causes, and 11.76% of murder. The cancer causes were diagnosed in hospitals in Sucumbios, Quito or Colombia.

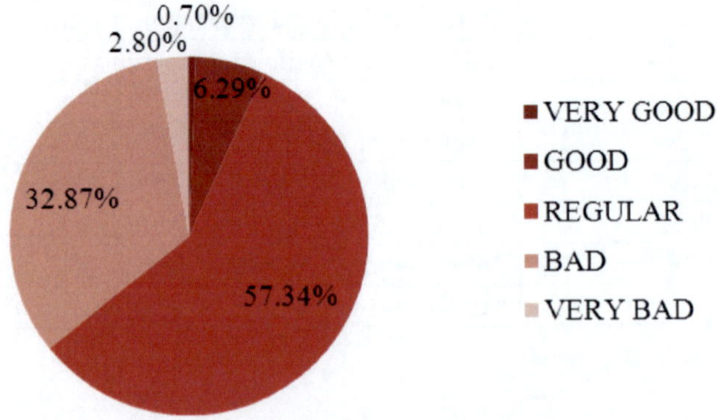

Figure 6.1. Personal health perceptions.

70.6% of the deaths occurred during the aerial spraying in connection with the *Plan Colombia* from 2000 to 2007, especially in 2003 and 2004 when most of the deaths occurred.

3. MATERNAL AND CHILD HEALTH

Of the total number of people surveyed and checked, 130 were women with a fertility record. Collectively, they had a total of 731 pregnancies. 44 of them (33.84%) had had their children before 2001 when the fumigations started, 39 of them (30%) only had children after 2000, and 49 of them (37.69%) had children before and after those dates. 73% of the children were born in 2001 while 27% were born after 2000. The median is between 5 and 6 pregnancies per woman, with a range of between 1 and 21 pregnancies. During the study six of these women were pregnant. The medical check-ups were conducted with a positive result, but some of them said they had previously had no medical attention at any time.

3.1. Abortions

Before 2001 the percentage of abortions was 8.43%, which increased up to 12.7% after the fumigations. At the same time, the percentage of dead children decreased from 12% to 9.14%, maintaining a tendency of life-loss from 80.15% to 77.66% (Tables 6.7 and 6.8).

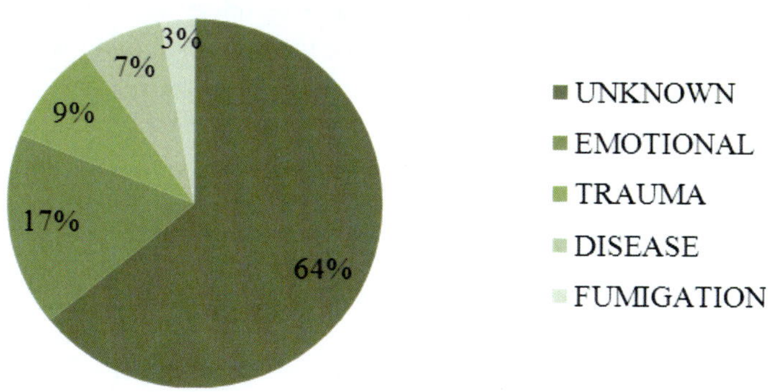

Figure 6.2. Causes of abortion.

Table 6.7. Fertility record before and after 2001

2001	Number of abortions	Number of deaths	Number of babies alive	Number of pregnancy cases
Before	45	64	428	534
After	25	18	153	197
Total	70	82	581	731

Table 6.8. Percentage of fertility record before and after 2001

2001	Abortions (%)	Deaths (%)	Babies alive (%)
Before	8.43	11.99	80.15
After	12.69	9.14	77.66
Total	9.58	11.22	79.48

The causes of these abortions are unknown in some 64.3%, emotional reasons (death of husband, fears, craving, domestic quarrels) in some 17.4%, trauma (blows, falls) in 8.57%, diseases (febrile, ectopia) in 7.14%, and the aerial spraying in connection with the *Plan Colombia* in 2.86%. This percentage is very similar to the increase detected in the surveys, but this percentage of the total accounts for 8% of the causes of abortion that occurred after 2000 (Figure 6.2).

3.2. Malformations

The family surveys revealed five cases of congenital malformations. Of these, 80% took place after the beginning of the aerial spraying, between 2001 and 2004. Only one of them happened before this period, in 1995 (Figure 6.3).

3.3. Child and Youth Mortality

The causes of deaths of the 82 children were very different. 17% of the causes are unknown. The primary cause of death (40%) is diseases (21.9% for diarrhea, 18.7% for fever from unknown origin, 15.6% for preventable diseases [tetanus 9.4% and whooping cough 6.2%], 12.5% for acute abdomen, and 12.5% for pneumonia, followed by meningitis, anemia, parasites, and cerebral palsy). The second cause of death is diseases linked to labor (podalic, premature) with children who died a little time after being born (13%). The third cause is death related to violence (9%); of these, 71.4% were murdered, 60% when they were underage (3 to 14 years old), 14.3% of deaths caused by violence concerned suicide, and the remaining 14.3% were caused by domestic violence against children. The number of suicides in this population is 191 people in every 100,000 people, which turns out to be excessively high since the WHO established 18 people for every 100,000 people as a standard. The fourth cause of death was witchcraft with some 6%, followed in fifth place by the aerial spraying carried out by the *Plan Colombia* with some 5%, and the sixth cause of death is leukemia or cancer with 4%.

3.4. Family Planning

Of the 130 women with a fertility record, 19% were no longer able to have children (menopause), 30% organized family planning, and only 51% had the possibility of becoming pregnant.

As for the women who organized family planning, 32.5% went for ligation of the Fallopian tubes, some 32.5% decided on intramuscular contraceptive, 25% oral contraceptive, and some 10% went for other types of methods (Copper T and calendar-based).

21.9% of them have been planning for less than a year, 46.9% for 1 to 3 years, 15.6% for 4 to 10 years, and 12.5% for more than 10 years.

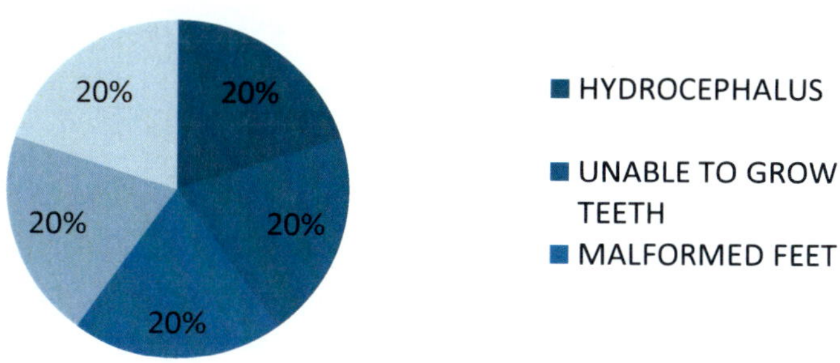

Figure 6.3. Classification of malformations in children.

4. Personal Record

4.1. Animal Bites

Of the total number of people checked, 98 (18.8%) affirmed to have been bitten by some animal (mammal and snake bites were only considered), of which dog bite was the most common bite (70.4%) followed by snakes (22.4%). Over time it was highlighted that there was a considerable increase in the percentage of dog bites, since 62.9% of them had taken place less than 10 years ago, in contrast with some 20.6% that had occurred between 10 and 20 years ago, and 16.5% that had happened more than 20 years ago. Nevertheless, concerning snake bites, there is no significant variation because 40% is maintained for the periods 0-10 years ago and 11-20 years ago, and 20% more than 20 years ago.

4.2. Acute Intoxication

Following the interviews with the individuals, it was observed that 9.4% answered affirmatively regarding some type of intoxication. Among this group, 57% claimed to have suffered from intoxication with pesticides, especially due to the use of Furadan, Gramoxone, and glyphosate; 6% had suffered from food poisoning, 2% from cocaine, and it is highlighted that some 29% of those affected claimed to have suffered severe intoxication due to the aerial spraying carried out by the *Plan Colombia*. This type of intoxication has occurred mainly over the last 10 years (with 77.1%), whereas 12.5% took place earlier, between 10 and 20 years ago, and a similar 10.4% occurred more than 20 years ago.

In order to have an idea about the extent of the use of agrotoxics, of the 144 families that were surveyed and checked, 84 reported having used agrotoxics. 58.3% of the families admitted to having used them more often, some 48.55% each month, some 33.62% over the last 6 months, 7.76% during the last 6 months to a year, whereas 6% used them three years ago, and some 6% more than three years ago. The frequency of use of the agrotoxics is detailed in the upcoming Table 6.9.

Of the total of the studied population (521 people checked), 239 are adults (45.9%), 118 of them had used agrotoxics, which accounts for 49.4% of the adult population. But also 11 underage people (from 11 to 17 years old) had used agrotoxics, which accounts for the 9.3% of those who had used them. An important piece of data is that more than one-third (36.44%) are women who handle chemicals both at home and on the farms.

4.3. Infectious Diseases

Almost one-third of the population (29.2%) stated having suffered from malaria, 6.7% had suffered diseases linked to water pollution (typhoid, hepatitis, EDA), which is a similar figure to those who have suffered from dengue fever, and respiratory problems, such as pneumonia and tuberculosis (TBC) are placed way behind.

4.4. Habits

4.4.1. Smoking

Smoking is a habit carried out occasionally by some 8.8% of the adult population in social meetings. Systematic or daily consumption only takes place among 7.9% of the adults. Of this population 63% have smoked only one to three cigarettes daily with a range from 1 to 40 years and a median of 16 years. 31.5% have smoked from four to ten cigarettes daily in a range from 8 to 40 years with an average of 18 years; whereas the remaining 5% have smoked twenty cigarettes for 40 years.

4.4.2. Drinking (Alcoholic beverages)

35.15% of those who admitted to drinking at social events were adults, whereas 4.75% of the habitual consumers were underage.

In addition to this, 1.67% of the adults said that they had drunk once or twice per week for 32 to 40 years. One individual from this group claimed to have quit drinking a year ago due to cirrhosis.

4.4.3. Drugs

Nobody admitted to having used any type of drugs. Only a single Indigenous person claimed to have used ayahuasca (a psychedelic brew of various plant infusions prepared with the *Banisteriopsis caapi* vine) for medical purposes, and a Colombian individual affirmed having used cocaine base 9 years ago.

Table 6.9. Distribution of the use of agrotoxics by family

Distribution of the use of agrotoxics by families					
Communities	Families YES agrotoxics		Families NO agrotoxics		Total
	#	%	#	%	
Chone 2	8	53.3%	7	46.7%	15
Yanamarum	3	33.3%	6	66.6%	9
Playera Oriental	2	18.2%	9	81.8%	11
Fuerzas Unidas	15	93.75%	1	6.25%	16
Puerto Escondido	13	92.8%	1	7.2%	14
Corazón Orense	13	72.2%	5	27.8%	18
Santa Marianita	5	41.6%	7	58.3%	12
San Francisco	6	30%	14	70%	20
Las Salinas	10	76.9%	3	23.1%	13
5 de Agosto	9	56.25%	7	43.75%	16

4.4.4. Medicine consumption

Several communities referred to the fact that the Ecuadorian Ministry of Public Health had visited them during the week in which the surveys were carried out; therefore, the rate of the use of medicine is not going to be accurate in accordance with what really happens in the communities. Upon this clarification, the type of medicine that the people were taking during the check-ups showed the following distribution: 17.7% of the population were taking medicine. Only one person (0.19%) admitted taking herbal medicine. Of the people who took medicine, 36.9% took vitamins, 27.2% took anti-parasitic drugs, 17.4% took antibiotics, 13% contraceptives, 6.5% analgesic, 5.4% antihypertensive, 3% anti-inflammatory and cold tablets, some 2% took antacids, and 1% treatment for cholesterol.

5. Impact of the Aerial Spraying on Health

5.1. Effects of the Aerial Spraying on the Population's Health

Considering that the pesticide that was sprayed entered into almost half of the houses, it is not surprising to discover that 84.73% of the families stated that some of their relatives had become sick during the aerial spraying.

The heads of households admitted in the surveys that 81.6% of them had suffered from some type of physical pathology during the aerial spraying, of which the most frequent effects were respiratory (19%), digestive (17.6%), eye problems (14.88%), central nervous system (14.88%), or skin problems (8.84%). After the aerial spraying, skin problems climbed to the first place (27.4%), and headaches maintained their prevalence (17.44%), while general discomfort decreased (8.19%), digestive problems (6.76%), fever (6.41%), and eye problems (4.63%) (Table 6.10).

84.6% of the population claimed to have had psychological effects, whereas 15.14% of those surveyed pointed out that they were calm and were not concerned at all. Of the psychological impact, fear was the most common reaction (51.3%), followed by concern (7.9%), nervous impulse (5.3%), sadness (5.3%), and curiosity (4.6%). It is noteworthy that after the aerial spraying, when the aircraft disappeared, fear decreased by 25%, but concern was doubled (18.6%), sadness increased three times (16.7%) when watching the impact of the aerial spraying, and nervous impulse was maintained by 5.8% (Table 6.10).

The feeling or perception of what could have happened to other relatives or family members is slightly different; consequently, during the aerial spraying, of those who gave the family surveys 27% did not identify any type of problems regarding their partners, 13.3% mentioned respiratory problems, digestive problems (12.1%), headache (11.5%), and skin problems (10.3%). After the aerial spraying, it was observed that skin problems considerably increased by three times (30.9%), the percentage of headaches was maintained (14.3%), and digestive problems (8.5%) and eye problems (5.8%) reduced (Table 6.11).

As for the psychological aspect, it was equally recognized that fear was the most common impact in almost half of the couples (48.31%), an impact that was reduced to 26.1% after the aerial spraying. Concern, which was not very significant during the aerial spraying (9.32%), increased after the fumigation by some 21%. Likewise, sadness tripled from 5.1% to 14.3%. In the meantime, despair was maintained at around 5% (Table 6.11). However, the witnesses did talk about serious effects after the aerial spraying.

The impact on children was also registered. During the fumigation, 26.3% of children did not suffer from physical impacts, but the main affectations were digestive (17%), respiratory (12.3%), dermatological (11.7%), headaches (11.1%), and eye problems (7.6%). After the aerial spraying, skin problems tripled (35.8%) and the incidence of headaches was maintained (9%) (Table 6.12).

Table 6.10. Impact during and after the aerial spraying

	Impact during aerial spraying				**Impact after aerial spraying**			
Surveyed	**Physical**		**Psychological**		**Physical**		**Psychological**	
	Impact	**%**	**Impact**	**%**	**Impact**	**%**	**Impact**	**%**
	Respiratory problems	19.07	Fear	51.32	Dermatological problems	27.4	Fear	25
	None	18.14	None	11.84	Headache	17.44	Concern	18.6
	Dizziness, faint, nausea	17.67	Concern	7.89	Discomfort	8.19	Sadness	16.7
	Eye problems	14.88	Nervous impulse	5.30	Fever	6.41	None	10.3
	Headache	14.88	Sadness	5.30	Nausea	8.19	Calm	7.1
	Skin problems	8.84	Curiosity	4.60	Fever	6.41	Nervous impulse	5.8
			Calm	3.30	Diarrhea	6.05	Despair	5.8
					Eye problems	4.63	Need	2.6
	Other	6.51	Other	10.45	Other	16.71	Other	8.1
Total		100		100		100		100

Table 6.11. Impact on couples during and after the aerial spraying

	Impact during aerial spraying				Impact after aerial spraying			
Surveyed	**Physical**		**Psychological**		**Physical**		**Psychological**	
	Impact	**%**	**Impact**	**%**	**Impact**	**%**	**Impact**	**%**
	None	27.3	Fear	48.31	Skin problems	30.9	Fear	26.1
	Respiratory problems	13.3	None	15.3	Headache	14.3	Concern	21.01
	Dizziness, faint, nausea	12.1	Concern	9.32	Dizziness, faint, nausea	8.5	Sadness	14.3
	Headache	11.5	Despair	5.9	Fever	7.2	None	12.6
	Eye problems	11.5	Curiosity	5.1	Eye problems	5.8	Despair	5
	Skin problems	10.3	Sadness	5.1	Diarrhea	5.4	Calm	4.2
					Respiratory problems	5.4		
	Others	14	Others	10.17	Others	22.5	Others	16.79
Total		100		100		100		100

Table 6.12. Impact on children during and after the aerial spraying

	Impact during aerial spraying				Impact after aerial spraying			
Surveyed	**Physical**		**Psychological**		**Physical**		**Psychological**	
	Impact	%	Impact	%	Impact	%	Impact	%
	None	26.3	Fear	49.6	Skin problems	35.8	Fear	28.2
	Dizziness, faint, nausea	17	None	24.4	Fever	11.2	None	23.1
	Respiratory problems	12.3	Crying	8.9	Diarrhea	11.2	Sadness	8.5
	Eye problems	11.7	Curiosity	7.3	Headache	9	Lack of energy	8.5
	Headache	11.1	Screaming	2.4	Dizziness, faint, nausea	7.8	Calm	6.8
	Skin problems	7.6			Stomach problems	6.3	Trauma	5.1
	Other	13.9	Other	7.4	Other	18.7	Other	19,.8
Total		100		100		100		100

The parents commented that the psychological impacts affected 49.6% of their children, and after the aerial spraying this rate was reduced to 28.2%. Likewise, 8.5% of the children in the communities were sad and only some 27.9% were calm or as if nothing had happened; nevertheless, the effects were noticeable on the children.

Most of the families reported that 2002 was the year when most of the impacts (cases of serious diseases) took place. 60.9% of the families suffered from one or more incidents of serious diseases.

60% of the individuals who received medical attention stated that the medical check-up had been useful, while 40% affirmed the opposite. Most of those with serious diseases were examined in private clinics or doctors' surgeries, followed by public health centers, such as provincial or national hospitals. The spread of these pathologies was not followed up by the health centers and the epidemiological alarms were not set off.

5.2. Types of Pathologies

During the different periods of aerial spraying the population commented that they had suffered from clinical manifestations that were registered by the doctors who carried out the medical examinations, and we have respected that. These manifestations refer to problems with:

Eyes: Vision loss, irritation (red eyes), tearing, and eye burning.

Digestive System: Nausea and feeling faint, lack of appetite and tiredness, diarrhea and fever.

Respiratory System: Suffocation, respiratory difficulties, dyspnea, and rhinopharyngeal secretion.

Skin: Pimples, itching, vitiligo, maculopapular rash.

Pregnancy: Abortion cases and a larger number of malformations were described.

Child mortality: The number of children who passed away following the aerial spraying increased.

Teeth: A case of periodontal abscess was described.

Locomotor System: Pain in joints and long bones is described.

Tiredness and weakness: Tiredness and weakness.

Central Nervous System: Headache, loss of consciousness, and dizziness.

Psychological manifestations: Insomnia, lack of energy, sadness, constant concern, anxiety, fear, anguish, psychosis, nightmare, aphasia. Possibly this type of pathology was one of the most serious and hidden manifestations that the population suffered.

During the community assemblies it was mentioned that both adults and children were concerned and afraid during the aerial spraying. The population was afraid of the diseases, but also the violence; they were afraid of the fact that the aerial spraying was happening again and its impacts. Children are still "frightened" and "hide every time they hear an aircraft", but adults also stopped sowing as they had done and affirmed themselves to be "psychologically ill".

6. Medical Check-Up

A medical check-up on a total number of 521 individuals was carried out. The results, described as follows, are an indicator of the health situation in the Northern border two years after the last aerial spraying in connection with the *Plan Colombia*. Of the total number of people checked, 312 (59.88%) reported having some type of physical ache, whereas 40.11% said they did not have any disease and were healthy. The most common aches found in the population had to do with the central nervous system, the locomotor system, the digestive system, skin, and the respiratory system.

6.1. Allergies

This type of pathology was found in 17 people (3.26%) who pointed out that they did not know the causes of their allergies. Almost one-third of them said that their allergies had come from medicine (penicillin, albendazole, and chloroquine), food (sardine, aloe vera), or dust. None of them had been thoroughly checked in order to find the most important causes of their disease. They had been diagnosed by doctors at that time without having received more than a medical warning.

6.2. Eyes

148 individuals, accounting for 28.4% of the analyzed population, had eye problems that mainly entailed pterygium (14.01%) and presbyopia (13.24%). In addition to that, 2.3% had viral or bacterial conjunctivitis, 0.96% had a

cataract, 0.77% had unspecified eye burning, 0.58% had trauma, 0.58% a sty, and a further 0.58% had strabismus.

It was also observed that 46 people had vision loss of more than 60% of the distance vision in at least one of their eyes, which means an affectation in some 10.84% of the population that went through this test. Out of all of them, 31 individuals (7.31%) had vision loss of more than 60% in both eyes, without having access to glasses. Consequently, if we include the cases of presbyopia (69) with serious vision loss (31) and deducting the coincidences of having both problems (17), almost one in every five people (19.57%) who live by the border needs glasses, which are indispensable for the 4% (17) who suffer from presbyopia and serious vision loss.

6.3. Digestive system

237 people, accounting for 45.48% of the analyzed population, were diagnosed with digestive diseases and problems classified as follows: parasitism (31.29%), gastritis and ulcer (9.41%), diarrhea (3.65%), vesicle stones (1.54%), as well as hernia (umbilical, inguinal, and hiatus), hemorrhoids (0.77%), and hepatitis (0.19%). The digestive problems are directly related to nutritional imbalance problems, poor nutrition, little or lack of hygiene on the part of food distributors, as well as food handling and preparation, causing parasites to be developed.

6.4. Skin

238 people (54.31% of the population) had skin diseases or problems which were divided into 33 different types of pathologies, as described below:

Diseases caused by fungi (39.2%): Tinea versicolor (15.5%), tinea pedis (9.98%), onychomycosis (7.49%), tinea corporis (4.41%), candidiasis (1.34%), seborrheic dermatitis (0.38%).

Pathologies caused by immunological alterations (6.525): Contact dermatitis (4.03%), lichen (1.73%), vitiligo (0.58%), eczema (0.19%).

Bacterial infections (6.33%): Piodermitis (4.99%), acne (0.77%), ganglion tuberculosis (0.38%), and pilonidal cyst (0.19%).

Hyperpigmentation pathologies (4.03%): Chloasma (3.07%) and nevus (0.96%).

Unspecific pathologies (2.69%): Rash (0.58%), prurigo (1.73%), granuloma (0.19%), and jaundice (0.19%).

Viral infections (2,1%): Wart (1.34%), cold-sore (0.38%), and exanthem (0.38%).

Parasite infection (1.34%): Scabies (0.76%), leishmaniasis (0.38%), and myiasis (0.19%).

Scars (1.15%): Scars (0.96%) and injuries caused after fumigations (0.19%).

Others: Injuries (1.15%), lipoma (0.76%), angioma (0.57%), and ulcer (0.19%).

6.5. Dentistry

Of the total population, 134 people (25.71%) had a perfect set of teeth, without cavities or loss of teeth. However, 55.27% had cavities or lacked one to ten teeth, whereas 10.17% had cavities or lacked eleven to twenty teeth. Also, 2.49% had 21 teeth with cavities and 6.33% had no teeth. One family had children with totally damaged teeth and they argued that it was due to medical causes, since they commented that the doctor had ordered them to take iron and that had caused them to damage their teeth.

6.6. Ears

A total of 98 people, accounting for 18.8% of the analyzed population, had some of the eight ear problems that are described as follows: inflammation (6.33%), fungal otitis (6.33%), hearing loss (3.45%), viral otitis (1.34%), bacterial otitis (1.15%), perforated eardrum (0.77%), polyps (0.19%), and foreign bodies in ears (0.19%). The ear problems were mainly identified in the communities that were directly related to really polluted water.

6.7. Nose and Throat

91 people (17.33%) were found to have some type of pathology, such as the following: catarrh (13.82%), tonsillitis (1.15%), nasal bridge deviation (0.58%), pharyngitis (0.38%), gingivitis (0.38%), herpangina (0.38%), saburral tongue (0.38%), and canker sore (0.19%).

6.8. Respiratory System

30 people (5.75%) were observed to have different types of respiratory problems: bronchitis (2.69%), tuberculosis (1.15%), asthma (0.96%), acute respiratory infection (0.58%), chronic obstructive pulmonary disease (0.98%), and sinusitis (0.19%). As for the cases of tuberculosis, 4 of them were already attended by the Ministry of Health and 2 were new cases. One of these cases was a family in which the mother who had this disease and her two children with cerebral palsy were being attended by the National Institute for Child and Families (INNFA).

6.9. Heart and Blood Vessels

63 people (12.09%) had some of the seven heart or blood vessel problems. The most common problem was arterial hypertension with 4.61%. Most of them commented they were not undergoing a treatment or that a doctor had given them a check-up a long time ago. 3.65% of the people who had received a check-up had hypotension, and 2.11% had varicose veins; in some of these cases surgery was needed. 0.77% had a murmur, 0.77% had arrhythmia, 0.38% had preeclampsia, and chest pain was found in 0.19% of the cases.

6.10. Kidney

Five kidney diseases were detected in 39 people (7.48%). Urinary tract infection was the most common (7.48%), prostatitis occurred in 2.81% of the male population, and kidney stones, pyelonephritis, and kidney atrophy each occurred in 0.19% of cases. The urinary tract infection exams were carried out with reactive strips that made it possible to detect on-site dysuria discomfort.

6.11. Locomotor System

Ten osteoarticular problems were found in 128 people, accounting for 24.56% of the analyzed population: arthralgia (10.75%), arthritis/osteoarthritis

(6.14%), lower back pain (5.18%), tendonitis (0.76%), traumas (0.57%), neuropathy (0.57%), spinal disc herniation (0.38%), kyphosis/scoliosis (0.38%), syndactyly (0.19%), and prominent sternum (0.19%).

6.12. Blood

Blood problems were found in 83 of the analyzed people (15.93%). The largest percentage of the five pathologies is for the people with anemia (14.59%). In order to get this diagnosis, the presence of three of these symptoms was analyzed: tiredness with little effort, waxen face, conjunctival pallor, delay of vascular return after pressure on nails, and mucus pallor. The cases of hypercholesterolemia (1.73%), hyperuricemia (0.58%), coagulation problems (0.19%), and dyslipidemia (0.19%) belong to results of recent blood analyses performed on participants in the check-ups.

6.13. Endocrine System

13 people (2.49%) were found to have four types of endocrine problems. 0.96% of those who were checked up had diabetes. These cases are treated with diet or tablets; insulin has not been necessary, but the control has been very poor. Metabolic syndrome has been found in 0.96% of the analyzed population, menstrual alterations in 0.58%, and hyperthyroidism in 0.19%.

6.14. Genitals

Of the 62 individuals (11.9%) who suffered from genital problems, 92% were women and 8% men. The most frequent pathologies in women were vaginal infections in 18.01% of the 249 women, followed by symptomatology that makes us suspect the presence of ovarian cysts (2.21%), fibroids (1.47%), mastitis (0.74%), hysterectomy (0.37%), and uterine prolapse (0.37%). In the 272 men the following pathologies were found: phimosis (1.2%), hydrocele (0.4%), and sexually transmitted diseases (STD) (0.4%).

6.15. Central Nervous System

Pathological problems of this kind were found in 132 individuals (25.33%). 10.36% affirmed they often had headaches followed by memory loss (0.96%), epilepsy (0.77%), fever seizure (0.58%), learning difficulty (0.58%), facial paralysis (0.38%), brain palsy (0.19%), and hydrocephalus (0.19%).

Regarding the medical check-ups, it may be concluded that 133 different types of pathologies were found and were distributed in a median of 2.82 pathologies per person (regardless of the dental afflictions which were the most common ones, such as cavities or teeth loss, which accounted for three-quarters of the population).

The most common diseases were dermatological and were found in more than half of the population, followed by digestive, eye, ear, nose and throat, and locomotor system problems (Table 6.13).

Table 6.13. Pathologies found in the medical examinations

Organs and systems	Number of affected people	Percentage (Total = 521)	Present diseases
Teeth	387	74.3%	2
Skin	283	54.31%	33
Digestive	237	45.48%	9
Eye	194	37.23%	9
Ear, nose, and throat	189	36.27%	17
Locomotor system	128	24.56%	10
Blood	83	15.93%	5
Psychological	68	13.05%	6
Central nervous system	64	12.28%	8
Heart	63	12.09%	7
Genitals	62	11.9%	9
Kidney and urinary tract	39	7.48%	5
Respiratory system	30	5.75%	6
Immunological system	17	3.26%	3
Endocrine system	13	2.49%	4
Total	1857	356.24%	133
Without dental problems	1470 = 2.82		

This classification makes it possible to learn what could be a common situation of diseases on the border that should be dealt with through the official processes in order to be able to somehow invest in better management and control of these diseases.

REFERENCES

Beristain, C; Fernández, I; Páez, D. Las palabras de la selva. Estudio psicosocial del impacto de las explotaciones petroleras de Texaco en las comunidades amazónicas de Ecuador. Bilbao, España. 2009.

Chevassus-Agnés, S. Población, pobreza y ambiente: Indicadores antropométricos de salud. *Food and Agricultural Organization (FAO)*., 1999.

Herrera, R. Estado de la nutrición en escuelas ecuatorianas de la frontera norte, afectadas por las aspersiones aéreas del plan Colombia. *Acción Ecológica*., 2006.

Maldonado, A; Saravia, A; Valladares, C; Álvarez, S; Bonilla, M; Maldonado, P; Supliguicha, V. Estudio de impacto de la actividad petrolera de Perenco, en el bloque 7 de la región amazónica ecuatoriana. *Acción Ecológica*. 2008.

World Health Organization (WHO). Medición del impacto nutricional: tablas. Ginebra, Suiza. 1979.

World Health Organization (WHO). The WHO, recommended classification of pesticides by hazard and guidelines to classification: 2004. *WHO IPCS*., Ginebra, Suiza. 2005.

Chapter 7

PSYCHOLOGICAL DIAGNOSIS

ABSTRACT

In this chapter the results obtained from the psychological analysis on girls and boys from eight Ecuadorian communities bordering Colombia are detailed. The psychological diagnosis was based on the analysis of the drawings made by the children at their respective schools. Several parameters that made it possible to draw conclusions about the children's emotional conditions caused by the aerial sprayings were analyzed. It was determined that in the communities there were lack of confidence, vulnerability, fear, lack of affection, repression, repressed aggressiveness, and depression. And finally it is important to make a development plan that makes it possible to overcome both the psychological and the social problems that the affected communities face.

1. PSYCHOLOGICAL TESTS

The psychodiagnosis of the drawings made by boys and girls from the following Ecuadorian communities bordering Colombia was carried out: Yanamarum, Chone 2, San Francisco, Fuerzas Unidas, Playera Oriental, Corazon Orense, Las Salinas, and Puerto Escondido. The purpose of this analysis was to get to know the current situation and the general context in which the everyday life of each one of these people takes place. The children at their schools were free to draw and paint. The analyzed parameters were the following: layout, extension, situation, openness, contrast, proportion, symmetry, support, concealment, confusion, aggressiveness, movement, rigidity, sketch, presentation, chromatic expression, outline, and texture. These parameters were useful for the interpretation and the conclusions at a psychological level (Table 7.1).

1.1. Yanamarum Community

Interpretation: On one of the drawings there is a helicopter flying over the crops where two unprotected people are standing. The vulnerability of this community is visible. There is an uncertainty of not seeing what is coming when there is direct contact with nature and where the possibility of having a quiet life is threatened by the presence of helicopters.

1.2. Chone 2 Community

Interpretation: On the analyzed drawings the repressed aggression is evident even though the adaptability to respond to the demands of the environment is perceived. There are also serious and deep conflicts under a rigid and unstable control. There are defeat and settlement feelings; the community is drifting away from reality with sadness and vulnerability. The features of aggressiveness, affective restriction, and anguish, as well as paranoid tendencies, hostility, and mistrust are found.

1.3. San Francisco Community

Interpretation: On the analyzed drawings the adaptation capacity to the environment is evident; nevertheless, there are features of lack of confidence, defenselessness, contact difficulties, as well as deep repression coming along with feelings of defeat, analytical capacity, energy, spontaneous expression, and originality. Additionally, repressed aggressiveness, paranoid tendencies, and mistrust are found.

What mainly calls the attention on many analyzed drawings is the presence of helicopters and other aircraft flying over the community that spill a special substance from the air that falls on animals, plants, and people.

Table 7.1. Parameters analyzed at the schools in Sucumbios

Parameters	Communities							
	Yanamarum	Chone 2	San Francisco	Fuerzas Unidas	Playera Oriental	Corazon Orense	Las Salinas	Puerto Escondido
Layout	Vertical	Vertical	Horizontal	Vertical	Vertical	Vertical	Vertical	Vertical
Extension	Spontaneous	Spontaneous	Spontaneous	Spontaneous	Spontaneous	Spontaneous	Spontaneous	Spontaneous
Situation	Adaptation Self-direction	Defeat Strength Serenity	Lack of self-confidence Depression Feeling defeated	Adaptation Self-direction	Adaptation Self-direction Over-watchfulness Introversion	Adaptation Self-direction Over-watchfulness Introversion	Adaptation Self-direction Over-watchfulness Introversion	Adaptation Self-direction
Openness	Yes	No	Yes	Yes	Yes	Yes	Yes	Yes
Contrast	Proportional	Disproportionate	Proportional	Proportional	Proportional	Proportional	Proportional	Proportional
Proportion	Yes	No	Yes	Yes	Yes	Yes	Yes	Yes
Symmetry	Balanced and analytical Capacity	No	Balanced and analytical Capacity	Balanced and analytical Capacity	Balanced and analytical Capacity	Balance	Balanced and analytical Capacity	Balanced and analytical Capacity
Support	Fantasy	Fantasy	Satisfactory Level and reality Fantasy	Satisfactory Level and reality Fantasy	Satisfactory Level and reality Fantasy	Satisfactory Level and reality	Satisfactory Level and reality Fantasy	Satisfactory Level and reality Fantasy
Concealment	No	Sadness	No	No	No	No	No	No
Confusion	Content difficulty	Content difficulty	Make it possible to identify its meaning	Make it possible to identify its meaning	The message of the drawings is determined	The message of the drawings is determined Blurry features	The message of the drawings is determined	The message of the drawings is determined
Aggressiveness	Unfriendliness features	Unfriendliness features	Found in few drawings	Unfriendliness features	Unfriendliness features	Unfriendliness features	Unfriendliness features	Unfriendliness features

Table 7.1. (Continued)

Parameters	Communities							
	Yanamarum	**Chone 2**	**San Francisco**	**Fuerzas Unidas**	**Playera Oriental**	**Corazon Orense**	**Las Salinas**	**Puerto Escondido**
Movement	Normal Adaptation	No	Normal Adaptation	Harshness situation Serious conflicts	Serious conflicts Little activity and initiative	Little activity and initiative	Normal Adaptation	Normal Adaptation
Rigidity	Vulnerability Rigidity	Vulnerability Rigidity	Vulnerability Rigidity	Vulnerability Rigidity	Vulnerability Rigidity	Vulnerability Rigidity	Vulnerability Rigidity	Vulnerability
Sketch	Energy	Aggressiveness	Energy	Energy	Energy	Aggressiveness	Energy	Energy
Presentation	Disorder Untidiness	Disorder Untidiness	Disorder	Order	Order	Tidiness	Tidiness	Tidiness
Chromatic expression	Uncontrolled emotion Spontaneous expression	Affective adaptability Affective restriction Distress	Affective adaptability	Affective adaptability Uncontrolled emotion Spontaneous expression	Affective adaptability Emotion Spontaneous expression	Affective adaptability	Affective adaptability Emotion Spontaneous expression	Balanced affective manifestations
Outline	Normal	No	Normal	Normal	Normal	Normal	Normal	Normal
Texture	Originality	Isolated features	Originality	Originality	Originality	Originality	Originality	Originality

1.4. Fuerzas Unidas Community

Interpretation: On the analyzed drawings a deep repression, expansion and aggression feelings, as well as rigidity are visible. There is aggressiveness along with paranoid defense, isolation, lack of confidence, the desire for protection and confidence, and a defenseless feeling. In addition to that, serious conflicts under a rigid and unstable control are evident. There is identification with the national symbols.

By means of the drawings helicopters are observed flying over the community whose people are standing in the same position in a defenseless and vulnerable manner.

1.5. Playera Oriental Community

Interpretation: On the analyzed drawings high repression is observed along with feelings of insignificance and inability to adapt, as well as rejection. Additionally, the need for protection, rigidity, and little control over the external world are observed. Lack of confidence is really noticeable.

The drawings show helicopters and aircraft flying over the community. The aircraft is spilling a substance. Thus, this fact makes it possible to confirm the interpretation of the drawings in which we can observe a lack of protection, and a lack of control over the environment, as well as a lack of confidence.

1.6. Corazon Orense Community

Interpretation: There are feelings of insignificance and inability to adapt, as well as a great sensation of rejection and rigidity. In addition to that, there is also an extreme repression with little control over the external world and a noticeable lack of confidence. An important need for protection is observed.

The drawings show helicopters and aircraft flying over the community and the crops. One of the drawings especially shows an aircraft and a boat with the national flag. In consequence, this confirms and emphasizes the extreme repression, the lack of control over the external world, and the lack of confidence, as well as the need for protection.

1.7. Las Salinas Community

Interpretation: The drawings show fantasy, repression, inability to adapt, rejection, and rigidity. Lack of control over the external world, paranoid tendencies, and watchfulness are observed. Additionally, there is a lack of confidence, and a need for affection and protection.

On the drawings we can see aircraft flying over the community and sprinkling a special liquid, which can be perceived as threatening; therefore, it allows us to verify the feelings of defenselessness and lack of confidence, as well as the need for affection and protection.

1.8. Puerto Escondido Community

Interpretation: Fantasy, a strong repression, rigidity and little control over the external world, along with lack of confidence, are observed. There are also signs of watchfulness and paranoid tendencies, as well as feelings of insignificance, inability to adapt, an extreme lack of confidence, the need for affection and protection, and defenselessness, which marks an important vulnerability.

The drawings show aircraft attacking a community and killing a human being.

Additionally, there are signs of spontaneity, adaptability to respond to the demands of the environment, expression capacity, originality, energy, expression, joy, strength, sensitivity, and creativity.

At the same time we can observe signs compatible with vulnerability, affective restriction, repression, anguish, inability to adapt, paranoid tendencies, watchfulness, contact difficulties, feelings of expansion and aggression, deep conflicts under an unstable and cautious control.

Bearing in mind what was previously described, the children from these communities show signs of lack of confidence, defenselessness, and inferiority, as well as a noticeable desire and need for protection and safeness.

2. Recommendations

- As a psychodiagnosis is available, it is important to carry out the respective follow-up in order to know the behavior that children develop during their social contact, so that their adaptation to the society and the environment is determined.

- Carrying out a psychological support is important since the evaluation made on the drawings shows the presence of aircraft and helicopters in a negative manner in the children's psyche because they drew aircraft opening fire against others, as well as aircraft spilling unknown substances from the air, which fell on the soil, human beings, animals, and plants.
- The results obtained from the current psychodiagnosis must be compared to the results from other previously done research, to determine whether the diagnoses were common or whether they demonstrate a progressive increase of fear, lack of confidence, and need for protection and safety, or whether the conditions have made it possible to diminish concern and fear, as well as anxiety.
- With the obtained results, it is important to develop a program, by means of days of multidisciplinary intervention, in order to promote individual, social, and community mental health, to mitigate the reactions of those emotions that at a certain time and under certain circumstances can act as barriers to children's psychosocial development.

Chapter 8

ENVIRONMENTAL DIAGNOSIS

ABSTRACT

This chapter discusses the environmental analysis made of the Ecuadorian communities that were exposed to the aerial sprayings with glyphosate. First, there is an explanation of the persistence capacity of the herbicide and its mobility in soils. The analysis of soils results in a high glyphosate and AMPA absorption capacity in clay soils found in the tropical humid forest. Additionally, there is an explanation of the physical, chemical, and biological indicators of the soil quality. On the other hand, this chapter focuses in detail on the results obtained from the analysis of the water quality in the bordering area, as well as the microbiological and parasitological analysis of water samples taken from the rivers flowing in the area of the communities exposed to the genotoxic.

1. ENVIRONMENTAL FATE FOR GLYPHOSATE AND ITS MOBILITY

The glyphosate chemistry is important to determine its fate in the environment. Glyphosate is physically and chemically similar to many substances found in nature and is not mainly reactive. It has been affirmed that glyphosate has little mobility in soils; it does not have a considerable biological persistence, it does not cause bioaccumulation and it does not biologically expand throughout the food chain (Giesy et al., 2000; WHO, 1994; Williams et al., 2000).

2. Mobility in Soils

Glyphosate has been manufactured to be directly applied to the plants' leaves. Although glyphosate is not directly applied to soils, a significant concentration of this compound may reach the soils (Haney et al., 1999).

Once it has reached the soils, there are different processes that determine the environmental fate of glyphosate:

- The complex formation of Ca^{2+} and Mg^{2+} ions found in water.
- The adsorption in sediments or particles suspended in water and soils.
- The entrance into the plant metabolism.
- Its biodegradation on the part of microorganisms.

The main regulatory process of mobility is when the herbicide is held back in soils. It has a high affinity when held back by the soil particles although there are previous experiences that prove there is loss by lixiviation through the preferential flowing pathways during rainfall after pulverization on humid soils.

The herbicide is degraded in soils due to mainly bacterial action in its metabolite, the AMPA, and the carbon dioxide (Rueppel et al., 1977; Sprankle et al., 1975; Mueller et al., 2003).

On the other hand, it has been found that the glyphosate adsorption varies according to the types of soils. There is less adsorption in soils with low contents of iron oxide (Piccolo et al., 1994). These authors have found that in some types of soils 80% of the herbicide that suffered from adsorption is released, whereas 15% and 35% of the herbicide is released in other soils.

The content of minerals in clay may also play an important role. A study conducted by Morillo et al. reveals that the glyphosate adsorption diminishes when copper is present due to the formation of glyphosate-copper complexes. This study concludes that it is necessary to consider the types of soils and the elements found in soils that are capable of forming complexes with glyphosate, in order to understand the relationship between the released glyphosate and its mobility in soils (Morillo et al., 1997).

Glyphosate can also join humus water-soluble substances. These substances are mainly responsible for the mobility of pesticides in soils. If glyphosate is transported by the humus substances, it can also reach the deepest soil layers (Piccolo et al., 1994). Welten determined that glyphosate can get attached to the soil particles and can even be toxic and bioavailable for organisms that ingest significant amounts of soil during their normal feeding (Welten, 2000).

It has also been discovered that the organic material of soils competes with glyphosate for adsorption sites (Gerritse et al., 1996). A Brazilian study conducted by Prata et al. shows that glyphosate competes with phosphorus for specific adsorption sites, especially in soils with a high phosphorus level (Prata et al., 2003).

3. Glyphosate Persistence

Another statement made in relation to glyphosate is that this herbicide is inactive and quickly degrades soils. The U.S. Environmental Protection Agency has reported that the glyphosate average lifespan in soils is up to 60 days (US EPA, 1999). In field studies residues are often found in the following year.

Some data on glyphosate persistence in different types of environment are detailed below (Cox, 1995):

- 249 days in farming soils in Finland.
- From 259 to 296 days in 8 forest areas in Finland.
- From 1 to 3 days in 11 forest areas in Sweden.
- 335 days in a forest area in Canada.
- 360 days in 3 forest areas in Canada.

Two Canadian studies found that glyphosate can persist from 12 to 60 days in a body of water after a direct application. Glyphosate residues were found in the sediments of a lagoon a year after a direct application in the United States.

Local studies in Canadian forests on glyphosate persistence in soils showed that glyphosate can persist from 45 to 60 days in such areas. After 360 days a presence of 6-18% of initial levels was found both in the soils and in the plant residues (Bell et al., 1997).

It has been found that the AMPA has more persistence than glyphosate. The average lifespan of this compound varies from 199 to 958 days (WHO, 1994).

Despite the fact that the effects of glyphosate, as a chemical product, have been researched in some types of soils, the effects of the surfactants and other additives used in the formulations for spraying activities in soils have not been researched.

It is clear that the conditions for glyphosate degradation depend on the climatic conditions in different areas. There is more persistence in cold weather or temperate-cold weather and there is less persistence in hot or tropical weather.

4. Description of the Area

The area where the research took place is located within the bio-weather called Tropical Humid Forest in the Amazon Lowlands Forest and in the zoo-geographic zone Eastern Tropical, identified as the representative ecological systems in the Amazon (Holdridge et al., 1971).

The soils have fine clay textures, muddy-clay textures with a prevalence of clay associated with a low degree of pedogenetic development or lack of cementing agents (organic materials) among its particles. Since the volume of the micro-pores and the capillary pores are greater, the soils have a good ability to hold back water. It is important to bear in mind that the prevalence of clay contributes to a greater retention of the ions in an exchangeable manner; therefore, they are more resistant to the lixiviation or percolation processes (Mejia, 1997).

The soils with a fixation of iron (Fe) and phosphorus (P) mainly include red soils, clay soils, and kaolinitic soils where phosphorus is the element that mostly has an influence on the color and is present like oxide or hydroxide (Mejia, 1997).

5. Soil Sampling

Soil sampling was carried out in accordance with the regulations of *The American Society for Testing and Materials* (ASTM), in order to establish the "Sampling Plan" of the research on soils by means of boreholes to take samples of soils.

The boreholes for the sampling met with the technical qualification by means of the application of the non-phosphate detergent and deionized water, prior to every extraction. In each community the sampling unit of 1 hectare was established. The subsamples were collected at some distance going over the area in a systematic zig-zag manner. 1 kg of the soil with duplicates was collected for quality assurance and quality control (QA/QC).

In order to guarantee the representation and the least possible alternation of the sample taken, when the local lithology makes it possible, soil sampling in the unsaturated area will be carried out. Considering the goals of the study, a direct push drilling equipment was used with the assistance of the Hollow Steam Auger.

The samples belonged to the soil profile A, in accordance with the goals of the study for the determination of glyphosate and AMAP residues. The samples were taken in native forests from 0 to 4 inches deep. As for the determination of the quality of soils by micro and macro-nutrient analysis, the sampling took place from 2 inches to 8 inches.

The samples were collected in paper bags and kept away in ziploc bags, codified, transported, and stored at 39.2° F for their analysis. Place, date, type of sample, depth, and room temperature were detailed.

5.1. Glyphosate Determination

The analytical methods to monitor glyphosate are based on classical chromatographic techniques, gas and liquid chromatography with precolumn or postcolumn derivatization. These techniques attached to mass spectrometry (MS) are also used. All these methodologies are expensive, use toxic solvents, and are highly specialized (Sancho et al., 1996; Seiber, 2002).

The analysis of glyphosate and AMPA residues in soil samples was determined by precolumn derivatization with FMOC at controlled pH. This derivate is injected in high performance liquid chromatography (HPLC) with detection by fluorescence. Before the extraction, 15 g of the sample was weighed for an alkaline pretreatment with KH_2PO_4 (monopotassium phosphate) at 0.1 M, the solution was shaken for 15 minutes at centrifugation of 3500 rpm for 10 minutes and filtration with filter paper Whatman No. 1. The extraction was repeated twice, obtaining an extract of 25 mL for each sample. The extracts were filtered on cellulose acetate membrane filter No. 0.45 μm. The derivatization consisted of adding 0.25 mL of borate buffer at 5% to 1 mL of the soil sample and 0.30 mL of 9-Fluorenylmethylchloroformate (FMOC) to 2 nM in a chloroform solution $CHCl_3$. The samples were placed at 104° F in darkness for 24 hours. Subsequently, 0.30 mL of H_3PO_4 (phosphoric acid) at 2% was added in order to stop the reaction, and the sample was placed in the refrigerator for the respective analysis. The derivatization product (Gly-FMOC) was analyzed by HPLC (Beckman, System Gold 126, detector 166 UV). The chromatography conditions are: Supelco RP-18 column (5 μm of a particle size, length X I.D: 10 inches x 10 nm); mobile phase: 0.05 M of phosphate buffer (pH 5.5); acetonitrile (65:35); current: 0.8 mL/min; and the detection with UV at 206 nm. The methodology was adapted and approved as specialized literature in technical-scientific analysis assessing the results of the study (APHA, 1998; Sancho et al., 1996; Stalikas et al., 2001; Veiga et al., 2001).

According to the results obtained in the glyphosate and AMPA analysis, these compounds were found with lower levels than the detection limit (0.05 mg/Kg) in the analyzed samples taken from the soil of the forests located as far as 820 feet away from the border with Colombia. This indicates that the clay characteristics of this soil have absorbed minimum residues of the glyphosate that came from the aerial sprayings.

5.2. Iron and Aluminum Determination

The national regulations and reference methods for the analysis of aluminum (Al) and iron (Fe) are not common; there are only some methodologies established at international level in order to characterize the presence of these atoms in soils and solid samples taken in polluted areas (US EPA, 1991). The analytical methods must be based on the recommended procedures for water, wastewater, sewage sludge, sediments, and farming soils.

The analytical technique of atomic absorption spectrometry (AAS) was applied for the analysis of the percentage of Fe and Al in the soil samples. By applying the reference method (Diaz-Romeu & Hunter, 1978), the compounds were obtained in different extracts, according to requirements, and their totality was obtained by acid digestion with classical methods.

Several environmental studies use data on the total presence of elements in soils or sediments although it is common to use this information together with other analyses made of extracts resulting from more solutions (citrate, dithionite, or ammonium oxalate) in order to learn about the potential environmental impact on the polluted soils (Holmgren et al., 1990).

In accordance with the analytical methodology, in the 18 samples the presence of Fe was between a minimum range of 1.0% and a maximum of 5.1%. As for Al, a minimum range of 1.5% and a maximum of 0.9% were registered in the studied communities.

Table 9.1. Analysis of glyphosate and AMPA in forest soil samples

Sample	Analite	Measured value (mg/Kg)
1 Forest June 22,2010	Glyphosate	< L.D.
	AMPA	D.N.C.
2 Forest June 24, 2010	Glyphosate	< L.D.
	AMPA	< L.D.
3 Forest June 26, 2010	Glyphosate	< L.D.
	AMPA	D.N.C.
4 Forest June 28, 2010	Glyphosate	< L.D.
	AMPA	D.N.C.
6 Forest June 27, 2010	Glyphosate	< L.D.
	AMPA	< L.D.
8 Forest June 27, 2010	Glyphosate	< L.D.
	AMPA	< L.D.
9 Forest June 22, 2010	Glyphosate	< L.D.
	AMPA	< L.D.

D.L. = Detection limit (0.05 mg/Kg); D.N.Q. = Detectable and non-quantifiable.

The results obtained from the primary forest soils show that there is an acid pH with prevalence of iron (1.0 % - 5.1 %) and aluminum (1.5% - 9.8 %) in the communities located in the studied area. On the Unified Environmental Legislation Text (TULAS, 2003) iron and aluminum are not included as parameters of environmental quality (LPCCA, 1976), despite the fact that they are heavy metals with a potential genotoxic and carcinogenic risk for human beings, plants, and animals. This implies a limiting condition for the quantification of glyphosate determination.

6. Farming Soil Quality

The Soil Science Society of America (SSSA) defines soil quality as "the functional ability of a specific type of soil to support animal or plant productivity, to maintain or to improve water and air quality, and to hold human settlement and health with natural ecosystem limits or those determined during its handling or use" (Karlen et al., 1997).

Soil quality includes the concepts of productive ability on the part of soils and environmental protection. The specific functions represented by soil quality include (Brejeda & Moorman, 2001):

- To get, to maintain, and to free nutrients and other chemical compounds.
- To get, to maintain, and to free water for plants and to refill the groundwater sources.
- To maintain an edaphic habitat adequate for soil biological activity.

Soil quality is dynamic and can change in the short term in accordance with its use and its handling activities. For its preservation it is necessary to implement sustainable activities in time (NRCS, 2004). Maintaining or improving soil quality can generate economic benefits in terms of production growth, better efficiency in the use of nutrients and pesticides, better air and water quality, and can reduce the effects of greenhouse gases (Brejeda, 2001). As previously affirmed, it is determined that soil quality is a dynamic property associated with the use of soil and its function, mainly for environmental production and livestock production.

6.1. Soil Quality Assessment

Soil quality assessment makes it possible to improve the response of the resources, such as: soil loss due to erosion, sediment deposits due to the wind or flooding, infiltration reduction and rain increase, hardening of the surface layer, nutrient loss, pesticide transportation, pH changes, increase of the availability of heavy metals, organic material loss, infestation of pathogenic organisms, and reduction of water quality (NRCS, 2004).

Due to the dynamism of soil quality, it can change in the short or long term in accordance with its use and its handling activities; therefore, it is necessary to monitor soil changes and to determine sustainable activities (NRCS, 2004). Selecting a set of soil quality indicators and developing their application in connection with a monitoring system are essential to assess soil quality.

The improvement of soil quality is generally perceived when the value of some characteristics increases. For instance, the infiltration or aeration rate may increase due to an increased amount of macropores, a larger size and stability of aggregates, and a larger quantity of organic material. But the supposed density, the resistance to farm work, the radical growth, the erosion rate, and the nutrient loss may decrease. A better assessment is achieved if, in addition to the aforementioned changes, other potential ecological-biological indicators are included, such as the degree of genetic diversity (both in the crops and the species of microorganisms, insects, and beneficial animals), the crop yield, the robustness of plants and their radical development, and the quality of the water that springs up the surface, as well as the water leached by underground lixiviation (Parr et al., 1992).

Maintaining or improving soil quality can generate economic benefits in terms of production growth, better efficiency in the use of nutrients and pesticides, and better air and water quality, and can reduce the effects of greenhouse gases (Brejeda & Moorman, 2001).

6.2. Soil Quality Indicators

Soil has no definite quality standards due to its variability; in consequence, it is very unlikely to establish a simple physical, chemical, or biological measure that adequately reflects it. Additionally, other factors that affect its efficiency must be considered, so that it is difficult to define, measure, and regulate soil quality (Bandinck & Dick, 1999).

Soil quality is assessed using indicators that show changes in soil capacity and its function (Dalurzo et al., 2002). The indicators depend on the ecosystem, so that characteristics that are useful as indicators of its sustainability should be determined.

The direct indictors that are commonly used belong to the soil's physical, chemical, and biological properties. Soil production capacity may be indirectly assessed with the efficiency of the farming, forest, and livestock systems.

According to the National Resource Conservation Soil (NRCS, 2004), soil quality indicators must meet the following conditions:

- To be easy to measure.
- To measure the changes of soil functions.
- To entail soil's physical, chemical, and biological properties.
- To be accessible to assessors and to be applicable in field conditions.
- To be sensitive to climate and handling variations.

Soil quality indicators allow analysis of the current situation and identification of critical points in regard to soil sustainability as a production means or as an important natural resource for life quality. They allow for the maintenance of biodiversity, analysis of the possible impacts before an intervention, monitoring the impact of the intervention, and help to determine whether the use of the resource is sustainable (Hünnemeyer et al., 1997).

The indicators can be qualitative variables (underground outcrop, canalicular erosion, puddle formation, etc.), quantitative variables (infiltration rate, cation-exchange capacity, pH, amount of nematodes or others), or indexes made up of the relationship between different variables (Astier-Calderón et al., 2002).

The evolution of soil quality can be determined in a comparative or relative manner. For such a purpose, the evolution of a system over time can be compared (longitudinal comparison) or one or more alternative handling systems with a reference can be simultaneously observed (transversal comparison) (Masera et al., 1992).

6.2.1. Physical indicators

There is an array of physical indicators of soil quality (Table 9.2). These vary in accordance with the prevailing characteristics of the studied area. Doran and Parkin selected the following as indicators: texture, depth, soil water infiltration rate, apparent density, and water-holding capacity (Doran & Parkin, 1994).

Chen et al. suggested that soil texture is related to porosity, water infiltration and availability; apparent density is associated with hydraulic infiltration and conductivity; and aggregate stability is related to erosion resistance and content of organic material.

6.2.2. Chemical indicators

The chemical indicators of oil quality include properties that affect the soil-plant relationship, water quality, shock-absorbing capacity of soil, and water and nutrient availability for plants and microorganisms (Table 9.2).

Doran and Parkin proposed the following as indicators: the content of organic material, or organic carbon and nitrogen, pH, electrical conductivity, ammonia nitrogen (NH_4+), phosphorus (P), and potassium (K). The indicators that reflect fertility standards (pH, MO, N, P, and K) are important factors in terms of crop production. Nevertheless, when the soil quality in polluted systems is assessed, other indicators are more important, such as the available trace elements: copper (Cu), zinc (Zn), cadmium (Cd), and lead (Pb), which were selected in Taiwan as chemical indicators (Chen et al., 2007).

Table 9.2. Soil quality: physical, chemical, and biological indicators

Indicator	Relationship to soil functions and conditions
Physical indicators	
Soil texture	Water and mineral retention and transportation; soil erosion.
Soil depth	Potential production and erosion estimation.
Apparent infiltration and density	Lixiviation, production, and erosion potential.
Water-holding capacity	Related to humidity content, transportation, and erosion.
Aggregate stability	Potential soil erosion, water infiltration.
Chemical indicators	
Organic material (organic C and N)	Soil fertility, stability, and erosion degree. Potential production.
pH	Chemical and biological activity, limits for plant and microbial activity growth.
Electrical conductivity	Microbiological and plant activity, limits for plant and microbiological activity growth.
Removable N, P, and K	Availability of nutrients for plants and potential loss of N, production and environmental quality indicators.
Cation-exchange capacity	Soil fertility and potential production.
Available heavy metals	Levels of toxicity for plant growth and crop quality.
Biological indicators	
Microbial biomass (C and N)	Potential microbial catalyst and C and N replacement.
N mineralized potential	Soil production and potential N contribution.
Edaphic breathing, water content, soil temperature	Microbial activity measuring.
Number of earthworms	Related to microbial activity.
Crop yield	Potential crop production, availability of nutrients.

6.2.3. Biological indicators

The biological indicators of soil quality generally have to do with the abundance and by-products of bacteria, fungus, nematodes, earthworms, annelids, and arthropods (Chen et al., 2007; Bautista et al., 2004).

Soil biological properties are dynamic because they are useful to provide early signs of soil degradation or improvement. Doran and Parkin selected the following as biological indicators: carbon and nitrogen from microbial biomass, potentially mineralized nitrogen, and edaphic breathing.

The biological and biochemical indicators (edaphic breathing, microbial biomass, enzymatic activities, microorganisms, and others) are essential in the interpretation of the dynamic of the organic material. They provide a quick response to the changes in the use of soil, they are sensitive to environmental stress, and they are easy to measure (Bandinck & Dick, 1999).

Table 9.3 shows the sensitivity and response of the most used biological indicators and the type of lab required for their measurement.

7. Indicators to Assess Soil Quality

The biomass of soil microorganisms accounts for at least 5% of its organic material; nevertheless, it plays at least three critical functions in soils and the environment: it is a labile source of carbon, nitrogen, and phosphorus; it is a reserve of these elements; and it is an agent of nutrient transformation and pesticide degradation (Dalal, 1998).

The comparison of the microbial biomass and indexes coming from it has been successfully used to measure changes induced by the handling practices, no farm work, crop rotation, cycle of nutrients, mud availability, or herbicide or pesticide application.

The current increase in the demand to monitor soil quality and to take care of the environment has caused techniques to become modernized to measure biomass or other equivalent indexes. The biomass has a fast change, which makes it experience very fast modifications in the face of environmental variations, appearance of pollutants, pesticides, etc., modifying its values before these changes appear in soil chemical or physical properties.

The measured parameters mostly found in the literature are: biomass, enzymatic activities like dehydrogenase, breathing activity, and also some indexes that combine two or more parameters, such as qCO_2 (breathing quotient) that belonging to n mg of CO_2 of breathing per hour and per kilogram of microbial biomass (Bautista et al., 2004).

Table 9.3. Sensitivity and response of the most used biological indicators and type of lab required for their measurement

Indicator	Sensitivity	Response	Determination
Carbon (%)	Very sensitive	Slow	Soil lab
Biomass C	Very sensitive	Fast	Specialized lab
Biomass N	Very sensitive	Fast	Specialized lab
Biomass P	Very sensitive	Fast	Specialized lab
Breathing activity	Very sensitive	Fast	Specialized lab
Specific breathing	Very sensitive	Fast	Specialized lab
Dehydrogenase activity	Very sensitive	Fast	Soil lab

7.1. Analytical Methodology and Results

7.1.1. Nitrogen

- The Kjeldahl method was applied.
- A high level with a range of 66.00 to 101.00 ppm was found in 6 communities, a mid level with a range of 46.00 to 51.00 ppm was found in 5 communities, and a low level with a range of 10.00 to 29.00 ppm was found in 7 communities (Figure 9.1).

7.1.2. Phosphorus

- The 2.5:25 Olsen Mod. extraction method ($NaHCO_3$ 0.5 N, disodium EDTA 0.01 M, and Superfloc 127) was applied for determination by molybdenum blue colorimetric $(NH_4)_2MoO_4$ and Sb and K double tartrate with unit mg/L.
- A high level with a range of 25.00 to 28.00 ppm was found in 2 communities; a mid level with a range of 9.00 to 19.00 ppm in 10 communities, and a low level with a range of 2.20 to 7.60 ppm in 6 communities (Figure 9.2).

7.1.3. Iron, manganese, and potassium

- The 2.5:25 Olsen Mod. extraction method ($NaHCO_3$ 0.5 N, disodium EDTA 0.01 M, and Superfloc 127) was applied for determination by atomic adsorption spectrometry with meq/L.

- As for iron (Fe), a high level with a range of 120.00 to 2,100.00 ppm was found in 18 communities (Figure 9.3).

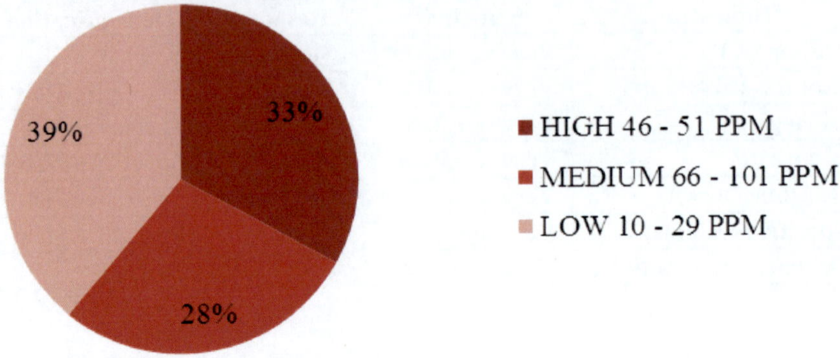

Figure 9.1. Nitrogen.

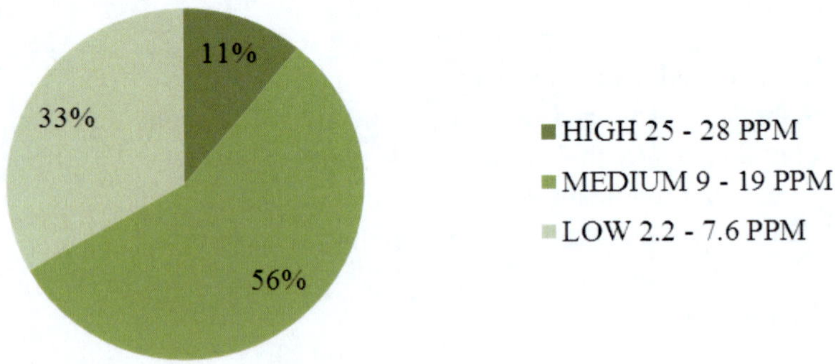

Figure 9.2. Phosphorus.

Figure 9.3. Iron.

- Concerning manganese (Mn), a mid level with a range of 5.9 to 14.6 ppm was found in 8 communities and a low level with a range of 2.8 to 4.6 ppm was found in 10 communities (Figure 9.4).
- Regarding potassium (K), a high level with a range of 0.52 meq/100 mL was found in 1 community, a mid level with a range of 0.20 to 0.37 meq/100 mL in 7 communities, and a low level with a range of 0.13 to 0.19 meq/100 mL in 10 communities (Figure 9.5).

7.1.4. Organic material

- The $K_2Cr_2O_7$ extraction method for 1Nf determination by 0.5 N $FeSO_4$ titration in percentage unit was applied.

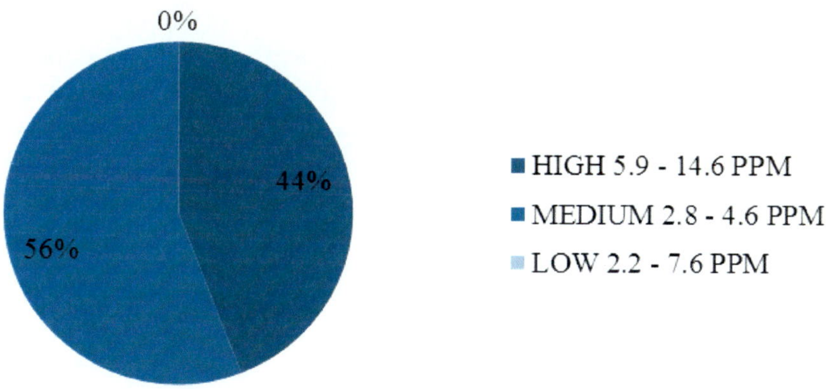

Figure 9.4. Manganese.

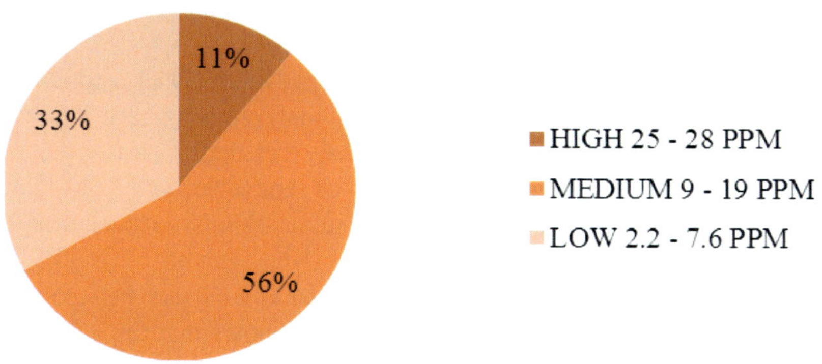

Figure 9.5. Potassium.

- A high level with a range of 4.20 to 6.00 % was found in 6 communities, a mid level with a range of 3.10 to 4.60 % was found in 8 communities, and a low level with a range of 0.90 to 2.60 % in 4 communities (Figure 9.6).

7.2. Determination of Macro and Micronutrients in Cultivated Soils

- The results obtained from the analysis of soils for nitrogen determination indicate that 33% of the samples have high levels, 28% have mid levels, and 39% of samples have low levels. Most of the communities have high and low nutritional levels; this interferes in plant growth and leaf coloration, and causes less fruit production.
- In relation to the quantification of phosphorus, it is observed that 11% of the samples show high levels, 56% of the samples have mid levels, and 33% of the samples have low levels. This means that half of the cultivated soils are in mid-level ranges, showing a correct plant ripening, ability for cell growth, division, expansion, and resistance to diseases.
- Iron is conclusive, having proven that all the lands in the studied area have 100% of iron. This concludes that this metal is involved in essential oxide-reduction reactions, hemoproteins, and non-heme proteins.
- The analysis of manganese shows that the soils are ranged from a mid level (44%) to a low level (56%). This means that the enzymatic processes related to soil nitrogen metabolism and carbonate decomposition have mid and low levels.
- Potassium registers a high level (5%), mid level (39%), and low level (56%). This indicates that most of the analyzed soils have a low level for this macronutrient, showing that the deficit may affect the photosynthesis process, activation of enzymes with biochemical functions, and the resistance to adverse conditions, such as drought or diseases.
- The organic material of the studied soils has a balance between a mid level (44%) and a low level (56%). This shows that the soils have a low and a mid percentage of nutrient availability for plant growth, the energy source for both macroorganisms and microfauna.

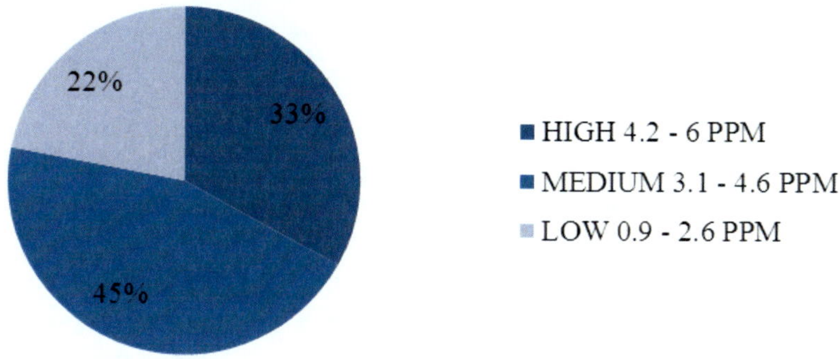

Figure 9.6. Organic material.

8. WATER QUALITY

The epidemiological profile in the bordering area is established by the poor availability of water safe for human consumption and the pollution of the water sources, affecting most of the people who have suffered from diarrhea and gastrointestinal diseases since the beginning of the aerial sprayings. There is a larger epidemiological risk of pollution of the water sources when the microorganisms are pathogenic (bacteria, virus, protozoan, and hookworms). This happens mainly due to poor sanitary practices in the community.

The microbiological and parasitological assessment makes it possible to analyze water quality, defining how acceptable water is for human consumption. This is the most common pollution directly and indirectly caused by wastewater, fecal excretion, etc.

9. MICROBIOLOGICAL AND PARASITOLOGICAL ANALYSIS

For the microbiological and parasitological analysis, 100 mL of river water was collected separately, dipping sterile recipients with screw cap 8 to 16 inches deep. The samples were flamed and sealed with parafilm, in accordance with the *Standard Methods for the Examination of Water and Wastewater* (APHA, 1998). The samples were codified in each site, placed in ziploc bags, transported, and stored at 39.2° F for their analysis. Location, date, type of sample, and room temperature were identified.

9.1. Microbiological Analysis

For the water microbiological determination, the total enterobacteriaceae, fungus, total coliforms, and fecal coliforms were recounted. The methodology consisted of inoculating 1 mL of water in the central part of the Petrifilm MR plate, avoiding air bubbles and applying the dispenser to homogenize the sample.

Two hundred plates were incubated for each family to be analyzed: 3M™ Petrifilm™ *Enterobacteriaceae Count Plates* at 95° F for 24 hours (+/- 2 hours). VRBG culture medium on laminated films. Colonies colored by the TTC indicator. Fungus: *The Petrifilm™ Yeast and Mold* (YM) incubated from 68 to 77° F for 3 to 5 days. Sabouraud culture medium on laminated films, chemical indicators found: Indicator for phosphatase. For recounting E. coli/coliforms, coliforms incubated at 95° F for 24 hours (+/- 2 hours) with VRBL culture medium modified on laminated films, presence of TTC indicator; and for E. coli, incubation at 95° F for 48 hours (+/- 2 hours).

9.2. Parasitological Analysis

Each one of the water samples (100mL) was centrifuged at 60 rpm for 1 to 3 minutes; the supernatant was discarded and the obtained sediments were used as samples for the parasitological analysis (Rodríguez, 2000), from which the smears or films were prepared. Subsequently, several observations were made using the following methods: direct, iron hematoxylin staining (heidenhain), and modified Giemsa (Maldonado et al., 2006). Micromorphometric measurement of the detected parasite structures and shapes was carried out.

10. Water Quality Determination

10.1. Microbiological Results

10.1.1. Count of enterobacteriaceae

The results obtained from the San Miguel River water showed a total recount of 2,428 enterobacteriaceae. This value is four times larger in contrast with the registration of the Putumayo River (542).

10.1.2. Count of fungi

The results of the San Miguel River water show a total recount of 70 fungi. This value is three times larger in contrast with the Putumayo River water (18).

10.1.3. Count of Escherichia coli

The results obtained from the San Miguel River water show a total recount of 129 E. coli. This value is doubled in contrast with that of the Putumayo River water (51).

10.1.4. Count of total coliforms

The results of the San Miguel River water show a recount of 643 total coliforms. This value is similar to that of the Putumayo River water (448).

10.2. Parasitological Results

Neither parasites nor rotifers were found in the analyzed water samples. 25% of diatoms were found in San Miguel River and Putumayo River. Some 5% of fungi were reported in San Miguel River, in contrast with 25% in Putumayo River. The bacteria were found in 35% of the samples taken from San Miguel River, unlike 65% in Putumayo River.

References

American Public Health Association. (APHA). Standard methods for the examination of water and wastewater. 20th edition. Washington DC, USA: *American Public Health Association (APHA), American Water Work Association (AWWA) and Water Environment Federation (WEF)*. 1998.

Astier-Calderón, M; Etchevers-Barra, J; Maass-Moreno, M. Derivación de indicadores de calidad de suelos en el contexto de la agricultura sustentable. *Agrociencia.*, 2002, 36, 605-620.

Bandinck, A; Dick, R. Field management effects on soil enzymes activities. *Soil Biol Biochem.*, 1999, 31(11), 1471-1479.

Bautista, C; Etchevers, B; del Castillo, R; Gutiérrez, C. La calidad de suelo y sus indicadores. *Ecosistemas.*, 2004.

Bell, F; Hawkins, J; Lautenschlager, R; Pitt, D; Ride, K; Wagner, R. Motor-manual, mechanical, and herbicide release affect early succession vegetation in northwestern Ontario. *Forest Chron.*, 1997, 73.

Brejeda, J; Moorman, T. Identification and interpretation of regional soil quality factors for the central high plains of the Midwestern USA. In: Stott, D; Mohtar, R; Steinhardt, GC (eds). *Sustaining the Global Farm.*, 2001, 535-540.

Chen, S & Liu, Y. Study on the photocatalytic degradation of glyphosate by TiO_2 photocatalyst. *Chemosphere.*, 2007, 67, 1010-1017.

Cox, C. Glyphosate, part 2: Human exposure and ecological effects. *J Pestic Reform.*, 1995, 15(4), 14-20.

Dalal, R. Soil microbial biomass - what the numbers really means? *Aust J Exp Agr.*, 1998, 38, 649-665.

Dalurzo, H; Serial, R; Vázquez, S; Ratto, S. Indicadores químicos y biológicos de calidad de suelos en oxisoles de Misiones, Argentina. *Facultad de Ciencias Agrarias-UNNE*. Misiones, Argentina. 2002.

Diaz-Romeu, R & Hunter, A. Metodología de suelos, análisis químicos de suelos y tejidos vegetal e investigación en invernadero. Turrialba, Costa Rica. 1978, 68.

Doran, J & Parkin, T. Defining and assessing soil quality. In: Doran, J; Coleman, D; Bezdicek, D; Stewart, B (eds). Defining and assessing soil quality for sustainable environment. *Soil Science Society of America.Special Publication 35.*, Madison, SA. 1994, 3-21.

Gerritse, R; Beltran, J; Hernandez, F. Adsorption of atrazine, simazine, and glyphosate in soils of the Gnangara Mound, Western Australia. *Aust J Soil Res.*, 1996, 34(4), 599-607.

Giesy, J; Dobson, S; Salomon, K. Ecotoxicological risk assessment for Roundup® herbicide. *Rev Environ Contam Toxicol.*, 2000, 167, 35-120.

Haney, R. Effect ofglyphosate on soilmicrobialactivity. *Proc Weed Sci Soc.*, 1999, 52, 215.

Holdridge, L; Grenke, W; Hatheway, W; Liang, T; Tosi, J. Forest, environments in tropical life zones: A pilot study. *Pergamon Press.* Oxford, UK. 1971.

Holmgren, P; Holmgren, N; Barnett, L. Index Herbariorum. *Ed. New York Botanical Garden.* New York, USA. 1990.

Hünnemeyer, J; De Camino, R; Muller, S. Análisis de desarrollo sostenible en centroamérica: Indicadores para la agricultura y los recursos naturales. *IICCA/GTZ.* San José, Costa Rica. 1997.

Karlen, D; Mausbach, M; Doran, J; Cline, R; Harris, R; Schuman, G. Soil quality: A concept, definition, and framework for evaluation. *Soil Sci Soc Am J.*, 1997, 61, 4-10.

Maldonado, A; Piedra, I; Maldonado, P; Bonilla, M; Chiriboga, A; Herdoíza, C; Rodríguez, E; Suárez, W; Herrera, R. Estado de la nutrición en escuelas ecuatorianas de la frontera norte, afectadas por las aspersiones aéreas del plan Colombia. *Acción Ecológica.*, 2006.

Masera, O; Arshad, M; Coen, G. Characterization of soil quality: physical and chemical criteria. *Am J Alternative Agric.*, 1992, 7, 25-31.

Mejia, L. Suelos del Ecuador. Reconocimiento general en base a su capacidad-fertilidad. Quito, Ecuador. 1997.

Morillo, E; Undabeytia, T; Maqueda, C. Adsorption of glyphosate on the clay mineral montmorillonite: effect of Cu (II) in solution and adsorbed on the mineral. *Environ Sci Technol.*, 1997, 31(12), 3588-3592.

Mueller, T; Massey, J; Hayes, R; Main, C; Stewart, C. Shikimate accumulates, in both glyphosate-sensitive and glyphosate-resistant horseweed (Conyzacanadensis L. Cronq.). *J Agric Food Chem.*, 2003, 51, 680-684.

National Resource Conservation Soil (NRCS). What is soil quality? (On line) Unites States Department of Agriculture. 2004 [10/05/2011]. Available from: URL: www.soils.usda.gov/sqi/soil_quality/what_is.

Parr, J; Papendick, R; Hornick, S; Meyer, R. Soil quality: Attributes and relationships, to alternative and sustainable agriculture. *Am J Alternative Agr.*, 1992, 7, 5-11.

Piccolo, A; Celano, G; Arienzo, M; Mirabella, A. Adsorption and desorption of glyphosate in some European soils. *J Environ Sci Health Part B.*, 1994, 29(6), 1105-1115.

Prata, F; Camponez, V; Lavorenti, A; Tornisielo, V; Borges, J. Glyphosate, sorption and desorption in soils with distinct phosphorus levels. *Scient Agric.*, 2003, 60(1), 175-180.

Rodríguez, M. Parásitos intestinales. Métodos de diagnóstico. Venezuela: *Escuela de Malariología y Saneamiento Ambiental "Dr. Arnoldo Gabaldeon"*. Maracay, Venezuela. 2000.

Rueppel, M; Brightwell, B; Schaefer, J; Marvel, J. Metabolism and degradation of glyphosate, in soil and water. *J Agric Food Chem.*,1977, 5, 517-528.

Sancho, J; Fernández, F; López, F; Hognedoom, E; Dijkman, E. Rapid determination, of glufosinate, glyphosate, and aminomethylphosphonic acid in environmental water samples using precolumnflurogenic labeling and coupled-column liquid chromography. *J Chromatogr.*, 1996, 737, 75-83.

Seiber, J. In pesticides in agriculture and the environment. *Wheeler BW, ed. Marcel Dekker Press*. New York, USA. 2002, 36.

Sprankle, P; Meggitt, W; Penner, D. Rapid inactivation of glyphosate in soils. *Weed Sci.*, 1975, 23, 224-228.

Stalikas, C; Konidari, C. Analytical methods to determine phosphoric and aminoacid group-containing pesticides. *J Chomatogr.*, 2001, 907, 1-9.

Texto Unificado de Legislación Ambiental Secundaria del Ministerio del Ambiente. (TULAS). Decreto Ejecutivo 3516 del Registro Oficial. 2003.

United States Environmental Protection Agency (US EPA). Description and sampling of contaminated soils. *A field pocket guide. Technology Transfer*. Ohio, USA. 1991.

Veiga, F; Zapata, J; Marcos, M; Alvarez, E. Dynamics of glyphosate and amino methyl phosphonic acid in a forest soil in Galicia northest Spain. *Sci Total Environ.*, 2001, 271, 135-144.

Welten, R. Ecotoxicity of contaminated suspended solids for filter feeders (Daphnia magna). *Arch Env Contam Tox.*, 2000, 9(3), 315-323.

Williams, G; Kroes, R; Munro, I. Safety evaluation and risk assessment of the herbicide Round up and its active ingredient, glyphosate, for humans. *Regulat Toxicol Pharmacol.*, 2000, 1, 117-165.

World Health Organization. (WHO). International program on chemical safety (WHO IPCS). Glyphosate. Geneva, Suiza. 1994, 159.

Chapter 9

MITIGATION AND REMEDIATION MECHANISMS

ABSTRACT

In this chapter the different useful mitigation and remediation mechanisms to deal with the different factors triggered by the exposure to genotoxic agents, such as glyphosate, are proposed. Plants as a source of antimutagenic agents, the conventional physiochemical and biological mechanisms, and the different enzymatic, and oxidation mechanisms are included among the topics discussed in the current chapter. The importance of applying useful mitigation and remediation mechanisms in the affected communities makes it possible to mitigate the development of problems in health and in the environment in the bordering areas that were exposed to carcinogenic agents, such as hydrocarbons or pesticides.

1. PLANTS AS A SOURCE OF ANTIMUTAGENIC AGENTS

The vascular plants have an enormous potential to create products that are useful in the chemical prevention of diseases related to mutations. Plants have been a source of useful bioproducts for both the treatment of diseases and the respective prevention. More than 120 bioproducts from vascular plants with a wide variety of therapeutic properties are currently obtained and it has been estimated that the products that come from these plants account for some 25% of the total amount of drug prescriptions in the Asian continent (Midorikawa et al., 2001).

Natural antimutagenic products in the diet are an important option as preventive chemical agents against cancer and other risky diseases, since most of the mutagenesis inhibitors that come from other sources can cause adverse effects (Deguchi et al., 2000). Natural plant products can be divided into terpenoids, alkaloids, phenolic compounds, and carbohydrates, presenting the first 3 anticarcinogenic agents thanks to their antimutagenic effect (Ferrer et al., 2002). As for the terpenoid group, phytosterol, carotenoid, monoterpene, sesquiterpene, and saponins are found. Carotenoids are good antioxidant agents that eliminate free radicals and reactive oxygen species (Arencibia et al., 2003).

A diet based on β-carotenoid, retinol, α-tocopherol or vitamin C is extremely beneficial for cancer reduction and longevity increase. Capsantine has been considered a better antioxidant than β-carotenoid and its capacity to eliminate free radicals is not influenced by esterification (Maoka et al., 2001).

D-limonene is a cyclic monoterpene found in the essential oils of numerous angiosperms, and it is known that it can act as a phase I enzyme inhibitor of the metabolism of xenobiotics and as an inhibitor of the post-transcriptional modifications of oncoproteins.

The anticarcinogenic effect that is attributed to the vitamin E (α-tocopherol), selenium, and the vitamin C (ascorbic acid) is the result of the antioxidant properties of these compounds. The most powerful antioxidant of the group of the vitamin E is α-tocopherol, which can act in a synergistic manner with β-carotenoid. Vitamin E is capable of functioning at high oxygen pressure and is a powerful degrading component of oxygen free radicals (Collins, 2001).

Alkaloids can be found in 20% of angiosperms and present analgesic, bactericidal, antibiotic, and psychotropic properties. Vinblastine and vincristine (isolated from *Catharanthus roseus*) used in chemotherapies against certain types of tumors are among the alkaloids that have pigments of monoterpenesin in their structure (Croteau et al., 2000).

Among the most studied compounds in order to seek chemical preventive agents are polyphenols that gather flavonoids (such as flavonols, flavonones, and isoflavonones), anthocianines, tannins, and lignans, much of which are found in fruit, plants, and very popular beverages, such as tea and wine. Polyphenols play an important role in nutrition and it has been stated that these compounds have antioxidant, antimutagenic, and anticarcinogenic properties. It is known that they inhibit the production of bisulfites in digestive organs, as well as protecting and stabilizing genome. Nevertheless, some of them turn out to be mutagenic and/or pro-oxidant since they interfere in essential biochemical pathways involving topoisomerases, prostanoids, and signal translation (Ferguson, 1998, 2001). Additionally, polyphenols such as cinnamaldehyde, coumarin, and tannic acid modify DNA replication and/or reparation. Some polyphenols can have a direct influence by modulating the genetic expression of genes that are part of error-free DNA repair (Abalea et al., 1999). The antioxidant activity in these compounds has been associated with its capacity to eliminate free radicals and to inhibit the P450 cytochromes or enzymes with oxidant activity, such as cyclooxygenases and lipoxygenases (Abalea et al., 1999).

Several experiments that have been conducted in multiple mutation assays identify tea as anticlastogenic and antimutagenic, attributing these properties to polyphenols that have been identified in the previously mentioned agents. Tea contains polyphenols, such as catechin, epicatechin, epigallocatechin, and epicatechin gallate, among others. It is important to highlight that green tea is very much used in cancer prevention since it inhibits the gene expression like the genes that codify for the tumor necrosis factor (TNF-α) and cytokine induced by TNF-α (Suganuma et al., 2000). In addition to tea, there are other plants that have a protective action, such as spinach and beet, from which it is known that the juice provides antigenotoxic activity in contrast with 2-acetylaminofluorene (Edenharder et al., 2009). Likewise, garlic has been considered an antioxidant compound (Imai et al., 1994). In a study in China it was pointed out that there is an inverse relationship between mortality and stomach cancer and garlic consumption, which proves the first evidence of its anticarcinogenic potential. Garlic and sulfurous compounds associated with it are reported as suppressors for tumor incidence in several body parts, including colon, breast, uterus, esophagus, and lung cancer (Shukla et al., 1999; Song & Milner, 1999). This protection is attributed to different mechanisms, such as the blocking of N-nitroso compounds, the suppression in the bioactivation of severe carcinogenic agents, the activation of the DNA repair, the reduction of cell proliferation, and the apoptosis induction (Arencibia et al., 2003).

Diet is the main source of human exposure to carcinogenic and environmental mutagenic agents, as well as anticarcinogenic and antimutagenic agents. Laboratory studies have identified several mutagenic and carcinogenic agents including polycyclic aromatic hydrocarbons (PAHs), heterocyclic amines (HAs), nitrosamines, mycotoxins, and plant carcinogenic compounds. There is a great deal of epidemiological evidence that a diet rich in fruit and vegetables can prevent human cancer in some way (DeMarini, 1998).

The organic sulfur compounds soluble in water are chemically preventive in the initiation stage of the carcinogenesis. The diallyl sulphides selectively inhibit some P450 enzymes and the development of carcinoma in colon, esophagus, and lung adenoma in rodents after exposure to carcinogenic agents.

Phytic acid (myo-inositol hexaphosphoric) is found in pollen and spores, as well as in leguminous plants, cereals, and inside the mammalian cells; in consequence, an important role of the phosphate groups of the phytic acid and the phosphate inositols is suggested in the protection from DNA oxidative damage (Midorikawa & Kawanishi, 2001).

2. Conventional Physicochemical and Biological Mechanisms

It is important to know the functioning of the different conventional and advanced mechanisms for the mitigation and remediation process of glyphosate on the varied nature surfaces exposed to the aerial and land spraying, so that the serious environmental impact caused by the herbicide concentration mixtures can be stopped in a certain way.

There are several studies that have made it possible to establish processes by means of which glyphosate is removed from water.

In the studies conducted by Hallas et al., the use of immobilized bacteria is proposed for the removal of glyphosate ($\leq$ 50 mg/L) from water and with low hydraulic retention time. They also developed the removal processes of glyphosate by using a reactor with a capacity of 45 L/min, achieving a degradation of more than 90% in 10 minutes of hydraulic retention (Hallas et al., 1992).

In the studies carried out by Speth, the use of activated carbon, conventional treatments, oxidation, filtration, and membrane separation techniques are proposed, in order to remove glyphosate from the drinking water sources in the affected communities. This mechanism was applied with low glyphosate concentration (0.796 mg/L) in distilled water and river water. The oxidation assays determined that glyphosate at low concentration is destroyed by chlorine and ozone. The oxidants hydrogen peroxide, hydrogen permanganate, and chlorine dioxide were not efficient for the removal of glyphosate. The use of alum coagulant determined that glyphosate is eliminated only if the water cloudiness is less than 0.2 NTU. The use of activated carbon determined that glyphosate is strongly absorbed in distilled water, but not as strong in torrential river water (Speth, 1993).

The research conducted by Ermakova et al. determined the efficiency of the bioremediation process in the soil affected by glyphosate concentration 10 times larger than the recommended amount. The glyphosate aerobic biodegradation was carried out from the use of the bacteria (families *Ochrobactrum anthropi* GPK 3 and *Achromobacter* sp. Kg 16), degradation ranges of which were around 2 to 3 times larger than the soil native bacteria, and the duration of which did not last longer than 2 weeks. It is proven that the use of bioremediation bacteria is an important strategy to mitigate the environmental impact caused by glyphosate aerial spraying or spill (Ermakova et al., 2010).

3. Enzymatic Mechanisms

The research done by Pizzul et al. consisted of studying the potential of the ligninolytic enzymes (peroxidase and lactase) to degrade glyphosate (10 mg/L) and formulations with glyphosate, such as the Roundup Bio. By means of the use of the ligninolytic enzymes, glyphosate was degraded within around 4 days. Despite the fact that the herbicide degradation was observed, there was no transformation in relation to the AMPA metabolite, by using the manganese peroxidase enzyme (Pizzul et al., 2009).

4. Oxidation Mechanism

The advanced oxidation processes (AOPs) and the advanced oxidation technologies (AOTs) have been useful in the glyphosate degradation found in the water. The AOPs use antioxidant agents (H_2O_2/Fe/UV, TiO_2/UV, UVC/O_3, H_2O_2/UVC) to generate highly oxidant species, such as the hydroxyl radical (·OH) that reacts with the herbicide, giving the green light to degradation. The advanced oxidation technologies have several advantages because they make it possible to degrade the polluting agent, generating its whole mineralization, they improve the organoleptic properties of the treated water, and they do not cause harmful side effects to health (Domènech, 2004).

Some of the advanced oxidation processes for the initial degradation of 42 mg/L of glyphosate in water are based on the use of the titanium dioxide catalyst (TiO_2) in suspension and UV radiation. This research proves that glyphosate can be easily degraded (Chen & Liu, 2007).

Huston and Pignatello studied glyphosate degradation in water through the photo-Fenton process. The degradation of active components in several pesticides was researched. The studied active principles were: aldicarb, azinphos-methyl, carbophuran, alachlor, captan, disulphoton, glyphosate, dicamba, metoxichlorine, malathion, picloram, and simazine. The results made it possible to observe that in most of the cases there was a complete loss of the active principles. Additionally, it was observed that the organophosphorade ingredients (glyphosate) proved to have less total organic carbon conversion for the same reaction time (Huston & Pignatello, 1999). The studies conducted by Chen et al. consist of glyphosate photodegradation in water through the ferrioxalate system ($Fe^{+3}/C_2O_4^{-2}$) under irradiation with a metal halide lamp. The glyphosate degradation (1 to 5 mg/L) turned out to be efficient. The researchers also propose a degradation mechanism whereby sarcosine and aminomethylphosphonic acid would be generated (Chen et al., 2007).

The process that uses UVC/H_2O_2 has been successful in the removal of several pollutants found in water and industrial effluent, including pesticides. The studies conducted by Hügul et al. have demonstrated that this process has enormous advantages due to the fact that the oxidant is commercially accessible, technically stable, has endless solubility in water, and requires a minimum investment; also, the operating process is easy (Hügul et al., 2000). The process that is carried out from the UVC/H_2O_2 has been applied by the *Universidad Nacional del Litoral* in order to degrade both water dilutions of glyphosate water and those of commercial formulations. In a study by Manassero et al. all the variants that affect the efficiency of this degradation process were analyzed, such as the pH, glyphosate concentrations, the H_2O_2, and the radiation levels (Manaserro et al., 2010; UNL, 2010).

The research group led by Balci et al. studied the capacity of the electro-Fenton process that was modified for the degradation of glyphosate water solutions, in which the Fe atom is replaced by Mn^{+2}. This process was carried out in an electrolytic cell with carbon cathode and platinum anode, and studied the effects of the applied current and the different Mn^{+2} and glyphosate concentrations that go from 17 to 68 mg/L. One of the most important remarks was the slower AMPA degradation (Balci et al., 2009).

REFERENCES

Abalea, V; Cillard, J; Dubos, M; Sergent, O; Cillard, P; More, l. Repair of iron-induced DNA oxidation by the flavonoid myricetin in primary rat hepatocyte cultures. *Free Rad Biol Med.*, 1999, 26, 1457-1466.

Arencibia, D; Díaz, D; López, Y; Rosario, L. Las plantas, fuente de agentes antimutagénicos y quimiopreventivos. *Rev Toxicol.*, 2003, 37-51.

Balci, B; Oturan, M; Oturan, N; Sirés, I. Decontamination of aqueous glyphosate (aminomethyl) phosphonic acid, and glufosinate solutions by Electro-Fenon-like process with Mn^{+2} as the catalyst. *J Agric Food Chem.*, 2009, 57, 4888-4894.

Chen, S & Liu, Y. Study on the photocatalytic degradation of Glyphosate by TiO_2 Photocatalyst. *Chemosphere.*, 2007, 67, 1010-1017.

Chen, Y; Wu, F; Lin, Y; Deng, N; Bazhin, N; Glebov, E. Photodegradation of glyphosate in the ferrioxalate system. *J Hazard Mater.*, 2007, 148, 360-365.

Collins, AR. Carotenoids and genomic stability. *Mut Res.*, 2001, 475, 21-28.

Croteau, R; Kutchan, H; Lewis, N. Natural products (secondary metabolites) chapter 24 from biochemistry and molecular biology of plants. *Amer Soc Plant Physiol.*, 2000, 12-948.

Deguchi, T; Ohba, R; Ueda, S. Radical scavenging activity of a purple pigment, hordeumin, from uncooked barley bran-fermented broth. *J Agric Food Chem.*, 2000, 48, 3198-3201.

DeMarini, D. Dietary interventions of human carcinogenesis. *Mut Res.*, 1998, 400, 457-465.

Domènech, X; Jardim, W; Litter, M. En: Eliminación de contaminantes por fotocatálisis heterogénea. Segunda edición. Editorial CIEMAT. Madrid, España. 2004, 6, 163-183.

Edenharder, R; Sanin, L; Carrasquilla, G; Solomon, K; Cole, D; Marshal, E. Regional differences in time to pregnancy among fertile women from five Colombian regions with different use of glyphosate. *J Toxicol Environ Health A.*, 2009, 72, 949-960.

Ermakova, I; Kiseleva, N; Shushkova, T; Zharikov, M; Zharikov, G; Leontievky, A. Bioremediation of glyphosate-contaminated soils. *Appl Microbiol Biotechnol.*, 2010, 88, 585-594.

Ferguson, L. Inhibitors of topoisomerse II enzymes: a unique group of environmental mutagens and carcinogens. *Mut Res.*, 1998, 400, 271-278.

Ferrer, M. Estudio del potencial antimutagénico de *Phyllanthusorbicularis*. Departamento de genética y microbiología. *Facultad de Ciencias. Universidad Autónoma de Barcelona (UAM).*, 2002, 1-91.

Hallas, L; Adams, W; Heitkamp, M. Glyphosate degradation by immobilized bacteria: fields studies with industrial wasterwater effluent. *Applied Environ Microb.*, 1992, 58, 1215-1219.

Hügul, M; Apak, R; Demirci, S. Modeling the kinetics of UV/hydrogen peroxide oxidation of some mono-, di-, and trichlorophenols. *J Hazard Mater.*, 2000, B77, 193-208.

Huston, P; Pignatello, J. Degradation of selected pesticides active ingredients and commercial formulations in water by the photo-assisted Fenton reactions. *Wat Res.*, 1999, 33, 1238-1246.

Imai, J; Ide, N; Nagae, S; Moriguchi, T; Matsuura, H; Itakura Y. Antioxidant and radical scavenging effects of aged garlic extract and its constituents. *Planta Med.*, 1994, 0, 417-420.

Manassero, A; Passalía, C; Negro, AC; Cassano, AE; Zalazar, CS. Glyphosate degradation in water employing the H2O2/UVC process. *Water Res.*, 2010, 44, 3875-3882.

Maoka, T; Mochida, K; Kozuka, M; Ito, Y; Fujiwara, Y; Hashimoto, K; Enjo, F; Ogata, M; Nobukuni, Y; Tokuda, H; Nishino, H. Cancer chemopreventive activity of carotenoids in the fruits of red paprika *Capsicum annuum*. *Cancer Lett.*, 2001, 172(2), 103-109.

Midorikawa, K; Kawanishi, S. Superoxide dismutases enhance H_2O_2-induced DNA damage and alter its site specificity. *FEBS Lett.*, 2001, 495, 187-190.

Pizzul, L; Castillo, M; Stenström, J. Degradation of glyphosate, and other pesticides by ligninolytic enzymes. *Biodegradation.*, 2009, 0, 751-759.

Speth, T. Glyphosate removal from drinking water. *J Environ Eng.*, 1993, 119, 1139-1157.

Suganuma, M; Sueoka, E; Sueoka, N; Okabe, S; Fujiki, H. Mechanisms of cancer, prevention by tea polyphenols based on inhibition of TNF expression. *BioFactors.*, 2000, 12, 1-6.

Universidad Nacional del Litoral (UNL). Informe acerca del grado de toxicidad del glifosato. Santa Fe, Argentina. 2010.

Chapter 10

CONCLUSION AND RECOMMENDATIONS

ABSTRACT

This chapter discusses the conclusions and recommendations obtained throughout the seven years of aerial spraying with glyphosate and two years of thorough analysis in such diverse fields as genetics, health, and environment on the part of a huge team of experts from the Biomedicine Research Institute in Quito, Ecuador. The different conclusions and discussion topics are based on the data obtained and analyzed from clinical backgrounds and surveys, medical examinations, analysis of detoxification genes and DNA repair genes, analysis of chromosomes, psychological tests, analysis of heavy metals, bacteria, and water and soil samples. Finally, the objective of this chapter - as well as this book - is to provide a baseline for further research conducted in the communities located in the province of Sucumbios that were affected by the glyphosate aerial spraying.

1. CONCLUSION

- From 2000 to 2007 the Ecuadorian territory bordering Colombia, especially the bordering communities of the provinces of Esmeraldas, Carchi, and Sucumbios, have been affected by the *Plan Colombia* which aimed at eradicating the production of illegal crops, such as cocaine (*Erythrozyllum coca*), opium poppies (*Papaver somniferum*), and marijuana (*Cannabis sativa*), by means of aerial spraying with a wide-spectrum herbicide. The program carried out by the Colombian

Antinarcotics Department (DIRAN-CNP) has caused political, social, and economic conflict between both nations.

- Due to the conflict generated because of the aerial spraying with highly concentrated herbicides, the Biomedicine Research Institute at the *Universidad de las Americas* carried out a project from 2008 to 2010 which focused on establishing a baseline in the genetic, cytogenetic, health, psychological, and environmental areas in 10 communities located in the province of Sucumbios that were affected by the aerial spraying.
- The research includes statements from 174 adults who lived in the 10 communities that were visited, the personal surveys of 144 parents (46.9% of the 307 registered families), medical examinations of 521 people (35.9% of the 1,450 residents), and the genetic and cytogenetic analysis of the peripheral blood samples taken from 182 individuals, some of whom were exposed to the aerial spraying and some of whom were not.
- The population is mainly Ecuadorian, originating from 17 provinces, but 16.1% of the families are from Colombia, a rate that has increased over the last 10 years since the beginning of the aerial spraying on illegal crops in Colombia.

1.1. Genetic Conditions

- In 2006 the DNA damage to 24 individuals exposed to the aerial spraying was evaluated through the comet assay technique which was widely used in studies with genotoxics, such as hydrocarbons, X-ray, and pesticides. The results showed a high level of DNA damage in the group that was exposed (comet length = 35.5 μm) in contrast with the control group (comet length= 25.94 μm), suggesting that those who were exposed to the wide-spectrum herbicide package had a genotoxic effect.
- Two years after the last aerial spraying it was observed that the total number of the studied population had no chromosome alterations, indicating their normal karyotype 46, XX; the percentage of chromosome fragility was found within the normal range.
- Regarding the genetic study, the GSTP1 gene codifies proteins which are involved in the xenobiotic metabolism and which play an important role in the regulation of the apoptosis process. A

statistically significant association of the allele belonging to valine in those exposed to the aerial spraying was observed (0.48), unlike in those who were not exposed (0.28).

- Peroxidase glutathione is one of the most important antioxidant enzymes in humans; it provides the oxidation in the DNA. The Pro198Leu polymorphism has been associated with lung, bladder, and breast cancer. Association of the allele belonging to leucine in those exposed to the herbicide was observed (0.41), unlike in the individuals who were not exposed (0.32).
- Protein codified by the XRCC1 gene is involved in the maintenance of DNA structural integrity, damage of which is induced by environmental factors. The Arg allele was more often found in the population exposed to glyphosate. The OR test determined that there is no significant risk in the population that has the Arg399Gln polymorphism.
- The genetic analysis made when the glyphosate aerial spraying took place in 2006 showed that the population had suffered DNA fragmentation, whereas the cytogenetic evaluation conducted two years after the last aerial spraying proved that the studied population had no chromosome alterations.

1.2. Social and Health Conditions

- The studied communities in the province of Sucumbios do not have good living conditions since they have no basic services. The sanitary and housing conditions are not the most adequate since their houses have inadequate ventilation and are built with zinc roofs, and 42.3% of them use no mosquito nets. Three-quarters of the families let in animals such as dogs and chickens. 77% of the households treat garbage, whereas 23% throw it away. The population, which initially lived as refugees, live in extremely poor conditions; therefore, they have been supported by the Ecuadorian people.
- Water consumption mostly depends on nature, since 38.8% of the supply comes from rivers, estuaries, or streams, whereas 25.2% comes from the rain. 21.58% of the water provision is from pits built by each family; in consequence, only 14.42% of the water is drinkable. As for feces elimination, only 38.4% of the families have some type of facilities, but 61.4% use the open field. Personal hygiene is important

and daily for 90% of the population, but 64.4% use soaps that are abrasive to the skin.

- 30.3% of the families own their lands, whereas 69.7% have possession of farms of around 21.3 hectares, although the range varies from one lot to 250 hectares, half of which are used for crops. This lack of property (proceedings and fee difficulties) makes people vulnerable and makes them an easier target for immigration, especially if their products are affected. The main activity is farming, mainly cacao, pasture, and coffee.
- More than half of the families (54.2%) use no agrotoxic products, and 76.56% of them do not use glyphosate. The use of glyphosate is not generalized or significant. 88.2% of the families that were surveyed do not use this chemical agent for their crops. Additionally, the use of this compound commonly takes place every six months.
- 81.6% of the population reported knowing nothing or little about the risks generated by the use of agrotoxic products (only 18.7% claimed to have enough knowledge of this issue). 64.5% of the people who use agrochemical products do not wear protection to avoid having contact with the agents: 73.8% do not wear a mask, 87.8% do not wear gloves, and 8.2% do not wear boots, but also 48.3% wash the clothes for fumigation together with the rest of the family's clothing. They admit that 53% of the fumigation equipment is not in good condition. 40% of them throw away the chemical containers in the environment and the 4.5% of those who handle agrotoxic products reuse the containers to collect or to drink water. This caused 9.4% of those who were examined to show intoxication due to pesticides.
- Concerning global nutrition, there is a noticeable improvement in the current nutritional condition in contrast with that of 2006, in the midst of the fumigation period. 2 years without fumigation have been enough for the global malnutrition condition in 6-17 year-old children to decrease from 10.3% to 3%; the risk of malnutrition also decreased from 36.3% to 23.2%. In terms of ages, the group of children under 6 years old has the highest malnutrition rate (14.9% for boys and 13.6% for girls). The malnutrition rate drastically decreases in the group of children from 6 to 11 years old (2% boys and 1.8% girls), whereas in the group from 12 to 17 years old the malnutrition rate increases to 3.8% for boys and 6.7% for women.
- As for chronic malnutrition, the current numbers are similar to those obtained in 2006 for the population from 6 to 17 years old; the

malnutrition rate went from 29% to 28% and the risk rate went from 35% to 40%. These numbers support the chronic role of the indicator and possibly more time is required to notice variations. By age groups, it is highlighted that the group with the worst malnutrition rate is the one from 12 to 17 years old (41%), possibly because this group already performs hard duties in farming production (such as the use of pesticides) and better nourishment would be necessary, in contrast with 30% of the group of children under 6 years old and 22% of those who are from 6 to 11 years old. It is also important to note that the malnutrition rate is higher in women (33%) than in men (24%).

- Regarding acute malnutrition, there are slight modifications in relation to the rates in 2006, going from an acute malnutrition of 1.87% to a current 1% and the risk rate went from 7.17% to 5.8%; nevertheless, it is highlighted that there is more obesity since the rate went from 5.9% to 16.3% which does not mean good nourishment, but a greater nutritional imbalance.
- The BMI in adults shows that after two years without fumigation those who are older than 18 years old have no malnutrition, but there is a tendency to obesity that affects 29.7% of women and 7.8% of men, and the rates increase with age. 8% of women claimed that the abortion they had suffered was due to the aerial spraying within the *Plan Colombia*.
- Concerning family health, the relevant data is that during the period of the aerial spraying within the *Plan Colombia* program, the abortion rate increased from 8.4% to 12.7%, whereas in the same period child mortality decreased from 12% to 9.1%.
- 80% of the malformations in the bordering area were affirmed to have happened after the beginning of the aerial spraying.
- The main mortality causes are: diseases (40%), unknown causes (17%), causes linked to giving birth (13%), violence (9%), evil eye (6%), the aerial spraying within the *Plan Colombia* program (5%), types of cancer (4%), transit accidents and congenital malformations (2%), and use of pesticides and snake bites (1%).
- The lack of warning that the fumigation activities were going to take place left people defenseless. 46.8% said that the chemical product got into their houses and 47.7% affirmed that they had continued with their activities during the aerial spraying. The communities were exposed to the impact of the aerial spraying because preventive

measures were not considered and the communities were not informed of the risks.

- 35.7% of the individuals stated that the substance that fell from the aerial spraying did not have the same shape, color, or odor, which makes us consider that an herbicide package mixed with several toxic compounds at high concentration was used during the spraying.
- In 84.7% of the families one of their members became sick during the aerial spraying due to the effects of the herbicides used. The symptoms were respiratory, digestive, ophthalmological, skin problems and headache, and after the aerial spraying the skin problems became the most important symptoms. Children were mostly affected by the aerial spraying. 32.4% of the families stated that during the fumigation a family member became seriously sick and this percentage increases up to 60.9% of the families after the aerial spraying.
- Most of the families said that the most visible impacts and the most serious diseases took place in 2002 and the medical attention was mainly provided in doctors' surgeries and private clinics; therefore, the epidemiological alarms were set off.
- Of the 521 people who were examined, 133 different types of diseases were found; especially the main affectations had to do with skin, digestive system, eyes, throat, nose, ears, and muscular and skeletal system. More than one-quarter of the population had these problems, and together with the problems found in the blood, the frequency of diseases was 76%.
- The clinical manifestations found in the population during and immediately after the aerial spraying were blindness, eye irritation, nausea, vomiting and dizziness, respiratory problems, rhinopharingeal secretion, maculopapular rash, pyodermitis, hair loss, intense itching, headache, loss of consciousness and fainting, insomnia, despondency, depression, continuous concern, anxiety, fear, anguish, psychosis, nightmares, periodontal abscess, pain in articulations, tiredness, weakness, increase of abortion and malformations, and increase of child mortality when the aerial spraying took place.
- The communities located in the bordering area that have been abandoned not only need better and more frequent medical attention, but they also need a boost on the part of the institutions that can rescue and offer better attention from health promoters, gynecologists, and herbalists, so that with kits with medicine, plants, and

acupuncture, the population can be offered first aid to prevent these diseases from becoming chronic and to treat those cases that need immediate identification, so that institutions such as the Provincial Board or the Ministry of Health itself can follow up the situation.

- In the case of chronic diseases, a lack of follow-up on the part of institutions and a lack of patients' will have been identified. It is necessary and very important to organize follow-up sessions and to gain the local populations' support.
- As for crops, the community testimonies and the family surveys agree that during the aerial spraying the short-cycle production and the cacao, coffee, and banana crops were drastically spoiled. Likewise, some plants died and fruit trees and the grass were affected.
- A loss of around 45.6% on farms, especially almost the whole grown lands, was mentioned; 39.8% affirmed the trees on the mountains also suffered damage.
- 96.3% of the families commented that, after the aerial spraying, the color, appearance, or flavor of plants changed. The families had to face the loss of their crops and the subsequent low production and loss or low fertility of their land. The characteristics of this damage that affected their production were diverse, but the effect was always the same: the peasants had no economic support from the sale of these products, and 96.1% of the families mentioned that food self-sufficiency was threatened. In addition to that, the rejection of the products that came from the bordering area caused the affected communities to suffer from hunger.
- 81.4% of the families affirmed that there was damage to the fertility of the land and it continues up to this day, affecting their sales volume and income. In search for solutions to this situation, the farming border of the communities has been expanded, by knocking down mountains and requesting new bank credits.
- 68.9% of the families said that the water was polluted and some people had to drink it because they had no other choice.
- As for the loss of animals, 37.9% of the families said they had seen dead fish floating on the rivers following the aerial spraying, but all of those who had fish tanks claimed to have had important losses during the aerial spraying. Likewise, they affirmed that there was a high number of abortion cases and deaths of cows, pigs, chickens, horses, dogs, and cats, as well as mountain animals such as monkeys, macaws, and parrots. 13.6% of the families admitted to having eaten

the dead animals in view of the fact that they were sure that nobody was going to recover their losses.

- In addition to the loss of animals due to disease or death, the sale of other animals in order to cover medical expenses was affected because the price was lower than the market price.
- The most intense period of aerial sprayings was between 2002 and 2004. Most of the cases involving serious diseases and dead animals were in 2002 as the highest mortality index took place in 2003 and 2004. The population mostly remembers the aerial sprayings that occurred in these years.
- Concerning the impact on the community, the aerial sprayings caused serious damage to the quality of life of those people who lived along the border, whose production and product commercialization were affected, along with their health, and their children's education, as well as the water and food quality, which remains the case today.
- The aerial sprayings were the main reason for 15.9% of the Ecuadorian families living in the border to move out to other places in the country in order to get away from the toxic effects of the herbicides; whereas some 14.6% of Colombian families arrived in the bordering communities having simultaneously fled from the violence and the aerial sprayings in the Colombian territory.

1.3. Psychological Conditions

- 84.86% of the families affirmed they had had some type of psychological disorders during the aerial sprayings, fear being the most frequent reaction with 51.3%. After the airplanes passed by, fear decreased and the concern for the future of the crops increased (18.6%), as well as unhappiness (16.7%).
- The drawings that were evaluated show different features: joy, spontaneity, adaptability, sensitivity, creativity, and the capacity for expression and adaptation to the demands of the media, as well as anguish, vulnerability, caution, paranoid tendencies, and watchfulness; the need for protection and safety is also obvious.
- The psychological manifestations and fear generated in children by the aerial sprayings are maintained up to this day. Some parents pointed out that there has been a behavior change in the children following the aerial sprayings with aggressive behavior, nervousness,

shyness, or a negative impact on their school attendance and performance. The psychological report conducted by means of children's drawings shows that the need for protection and safety is obvious. The expert advises psychological support for these children because the evaluation shows that the presence of airplanes and helicopters negatively affect the children's psychology since in many drawings these aircrafts are seen to open fire and in other drawings it can be seen that they spill a substance in the air which falls on human beings, animals, and plants.

1.4. Environmental Conditions

- Glyphosate is a little movable in soils, although this property must be understood in terms of the particular composition of soils. Nevertheless, there is scarce information or production of data in our country regarding this matter.
- The persistence of glyphosate in soils noticeably varies from weeks to few months, in accordance with warm or hot weather, and depending on the characteristics of the soil.
- There is a lack of regional studies on the persistence and mobility in soils; therefore, it is important to continuously monitor the bordering zone.
- In forest soil samples the presence of glyphosate and AMPA was determined in amounts that were lower than the detection limit. Since the areas in which the samples were taken are forest rather than farming zones, it is concluded that the geological characteristics of this soil absorbed glyphosate and only minimal amounts of the pesticide from the aerial sprayings remained.
- The results obtained from the analysis of the studied soil to determine nitrogen illustrate that 33% of the soils have a high level, 28% mid level, and 39% have a low level of nitrogen. The presence of high and low nutritional levels interferes in the growth of plants and the coloring of leaves, and delays and decreases fruit production.
- Regarding the quantification of phosphorus, 11% of samples show a high level, 56% a mid level, and 33% a low level, which means that half of the soils used for farming are at a mid level, whereby plants show the right ripening, as well as an easy cell growth, division, and expansion.

- All of the soils in the studied area contain 100% of iron, confirming that this metal is interfering in the essential oxide-reduction reaction, in hemoproteins, and non-hemic proteins.
- The analysis of manganese shows that the soils in the area are distributed from a mid level of 44% to a low level of 56%. This means that the enzymatic processes related to the metabolism of nitrogen and the decomposition of carbonates in the soils have mid and low levels.
- Potassium is registered at a high level (5%), mid level (39%), and low level (56%). This indicates that most of the soils that were analyzed have a low level of potassium, which means that the deficit may affect the photosynthesis process, the activation of enzymes with biochemical functions, and the resistance to adverse conditions, such as drought or the presence of diseases.
- The organic material of the studied soils is balanced between the mid level (44%) and the low level (56%). This means that the soils have a low and mid percentage of nutrient availability for plant growth, the energy source for both macroorganisms and microfauna.

2. Suggestions

2.1. Suggestions from the Communities

- To have a real presence in the area to enable the population's real needs to become known, since the local authorities only stay in the cities and do not often visit the communities, so they do not know the people and their problems.
- To support farming projects for boosting the organic production and the commercialization of their products by means of storage centers.
- To promote an inter-ministry coordination project along the border in order to face the problems and to provide solutions that are not only military in nature but also social, environmental, and health-related.
- To promote infrastructure works: roads, electricity, health care, water, toilets, and elementary and high schools.
- To promote programs to recover forests and the native fauna.
- To firmly keep rejecting the aerial sprayings and, consequently, to encourage the population to recover their hope of working and living in the affected areas.

- To improve roads and electricity, and to ensure the provision of drinking water which is as healthy as possible for the communities.
- To carry out projects to repair the psychological damage to children along the bordering area.

GLOSSARY

Chromosome aberrations: Abnormal chromosome number and structure.

Adsorption (in surface and colloid- chemistry): A process by which a component is transferred through two phases in contact and ends up being integrated in one of them.

Deoxyribonucleic acid (DNA): Complex molecule that is part of the chromosomes that store the hereditary information in variations in the sequence of the purine and pyrimidine bases. This information turns into protein synthesis; therefore, it is decisive for all the physical and functional characteristics of the cells and the organism.

Ribonucleic acid (RNA): Lineal polymer of ribonucleotides, each one of them has ribose sugar associated with a phosphate group and one of the four nitrogenated bases: adenine, guanine, cytosine, and uracil; it codifies the information for the sequence of the amino acids in protein synthesis.

Adenocarcinoma: Benign tumor developed in the glandular epithelium or forming structures of the glandular type.

Alkaloid: Compound of vegetal origin with one or more nitrogen atoms that make it have an organic base. There are also synthetics and those produced from proteins.

Allele: Each one of the different forms of a gene that appear in the same relative position (locus) of other counterpart chromosomes.

Environment: Whatever is around; a set of all the external conditions and influences to which the system to be studied is exposed.

Gene amplification: Production of copies of a sequence of an intra or extra chromosome DNA; as for plasmids, it refers to the increase of copies of the plasmid per cell, induced by a specific treatment of transformed cells.

Apoptosis: Physiological process whereby tissues die and disintegrate within the normal development of the living beings.

Suffocation: Situation that happens as a result of lack of oxygen absorption.

Asthma: Chronic respiratory disease that is characterized by bronchoconstriction, excessive mucus production, and edema in the lung alveoli caused due to respiratory difficulties, panting, and coughing.

Biomarker: Measures taken at molecular, biochemical, or cell levels in human populations, plants, or animals that come from polluted areas, as well as in organisms exposed to pollutants, and that affirm the organism has been exposed to toxic substances and the magnitude of the organism's reaction to the pollutant.

Cancer: Name of malignant tumors. Carcinoma begins in the epithelial cells; sarcoma begins in the conjunctive tissue.

Carcinogenesis: Progressive transformation process from normal cells to malignant cells. The acquisition of autonomy by themselves is produced, which reflects a regulation and abnormal expression of their genetic makeup. This process may be the result of endogen events like errors in the DNA replication, the intrinsic instability of some DNA bases, or the attack of free radicals generated during the cell metabolism. This can also be the result of exogenous processes such as ionizing radiation, UV radiations, and chemical carcinogens.

Carcinogen: Physical, chemical, or biological agents that are capable of increasing the incidence of malignant neoplasia.

Carcinoma: Malignant tumor of epithelial cells.

Case-control, study: Study that starts with the identification of individuals with a determined disease of interest, and an adequate health control group, the relationship of a characteristic with the disease by means of a comparison between the sick group and the healthy one, regarding the presence or number of characteristics in both groups.

Cytogenetic: Branch of Genetics that relates the chromosome structure and number in isolated cells with the variation of phenotype and genotype.

Coadjuvant: In Pharmacology, the substance that is added to a medicine to speed up or increase the activation of the main component. In Immunology, the substance or organism that increases the response of an antigen.

Community: In Ecology, set of communities that live in the same geographic area and have connections with one another.

Lethal concentration: Proportion of a toxic substance in an environment that causes the victims to die after some period of exposure.

Trust interval: A set of ordered values whereby the value of a parameter of a population is found with a probability that is determined by a level of pre-established trust. It measures the accuracy of the parameter estimation.

Conjunctivitis: Conjunctival inflammation.

Pollutant: Microbial agent unduly found in an environment. Minor foreign body found in a substance. Foreign material inadvertently added to a sample before or during the chemical or biological analysis. Undesirable component found in food, medicine, or any other product that may be risky to the consumer. Undesirable components found in the environment.

Control group: Selected or necessarily established group before a study, made up of human beings, animals, cells identical to the studied group, and that is maintained in the same situation and conditions without exposing it to any agent.

Chromatid: Each one of the two filaments that are stuck together by the centromere that forms a chromosome.

Chromatin: Coloring complex of DNA and proteins found in the nucleus of a eukaryotic cell.

Chromosome: Self-replicant structure formed by DNA associated with proteins involved in the storage and transfer of the genetic information; the physical structure containing the genes.

Dermatitis: Skin rash that normally occurs in a localized area, as a consequence of direct exposure to an irritating or infectious substance.

Dehydrogenase: Enzyme that catalyzes the oxidation of compounds through subtraction of hydrogen atoms.

Waste: Anything that is deliberately thrown away or that is used for different purposes in contrast with the first use.

Denaturalization: Changes in the molecular structure of proteins that avoid their normal functions; it is normally generated due to the alterations of the intramolecular hydrogen connections as a result of reactive substances or heat.

Dyspnea: Difficult and interrupted breathing.

Dose: Amount of a substance that is taken or absorbed by an individual in proportion to his/her weight or body mass, usually in 24 hours.

Average lethal dose (LD_{50}): A statistically calculated dose of a chemical or physical (radiation) agent that is expected to kill 50% of the organisms of a community under a set of determined conditions.

Ecology: Branch of Biology that studies the interaction between living beings and all the facts surrounded in the environment, including the other organisms.

Disease: Pathological situation gathering a set of peculiar symptoms that distinguish it as an abnormal entity in contrast with other normal or pathological situations.

Enzyme: Catalyst of the biochemical reactions, making the transformation of subtracts easier.

Species: In Biology, a group of organisms with common ancestors that is capable of reproducing among themselves, providing fertile descendants. In Chemistry, a pure chemical substance.

Exposure: A situation whereby a substance may affect, through any way, a community, organism, organ, tissue, or target cell.

Pharynx: Part of the digestive tube located underneath the esophagus and above the mouth.

Phenotype: Observable structural and functional characteristics of an organism determined by the genotyping and modulated by the environment.

Fertilizer: Substance used in agriculture to improve the nutrition of plants and to increase the performance of crops and/or to control production.

Gene: Basic structural and functional unit of hereditary material: an ordered sequence of nucleotide that codifies the synthesis of a polypeptide chain or a regulating sequence that makes the translation possible.

Genotype: Specific allele composition of a cell referring to the whole genome or more commonly to a gene or a set of genes.

Genotoxic: Capacity for the physical, chemical, or biological elements to generate alterations in the genetic material due to changes in the DNA or in the intracellular structures linked to the functioning or properties of the chromosomes. The genotoxic substances can be directly connected with the DNA or indirectly act by means of the affectation of the enzymes involved in the DNA replication, consequently causing mutations that may or may not result in cancer. The genotoxic substances are not necessarily cancerogenic, but most of the cancer-causing agents are genotoxic.

Genotoxicity: Capacity to cause damage to the genetic material; the damage may be mutagenic or cancerogenic.

Herbicide: Substance to eradicate plants.

Hygiene: Science of health and its conservation.

Incidence: Number of cases in which a disease is beginning, or in which people become sick, during a determined period in a specific population. It is usually given in percentage form, of which the denominator is the average number of people during said period, or the estimated number of people during half the period. The incidence refers to new cases, whereas the word prevalence focuses on all the new and old cases.

Inhalation: Air, steam, gas, or particles found in these elements which come in the airway.

Immune system: A set of organs, cells, vessels, and lymphatic glands that are part of the formation, activation, storage, and distribution of antibodies and mediators of hypersensitivity reactions.

Trust interval: A range of values of the variable that is being measured, which is mostly likely to contain the real value of the average or the proportion in the community or in the sample.

Intoxication: A pathological process with clinical signs and symptoms caused by an exogenous or endogenous substance.

Irritant: Substance that causes inflammation after immediate, prolonged, or repeated contact in skin, mucus, or other tissue.

Larynx: Main organ of the voice that begins with the respiratory tract, located from the larynx and the trachea.

Lymphocyte: Type of immune cell is elaborated in the spinal cord and is found in blood and lymphatic tissue. The two types of lymphocytes are the lymphocyte B and the lymphocyte T. The first type elaborates antibodies and the second type helps to destroy the tumor cells and helps to control the immune responses.

Miosis: Process of cell division which is normal with diploid cells, by means of which each daughter nucleus receives half the number of characteristic chromosomes of the somatic cells of the species. The result is gamete in animals and spores in plants.

Metabolism: All of the chemical and physical processes that take place in an organism; in a strict sense, physical and chemical changes that a substance in an organism experiences. It includes the incorporation and distribution in the organism of the chemical components, the changes and the elimination of the compounds and their metabolites.

Metastasis: Transfer of bacteria or other cells, especially cancerous ones, from one part of the body to another one, causing modifications in the space localization of a disease or its symptoms. It is the growth of pathogenic microorganisms or abnormal cells far from the original site in the body.

Mitochondria: Organule of the eukaryotic cells surrounded by an external and an internal membrane. The internal membrane has folds called crests in which the ATP synthesis takes place in the oxidative phospholarization in the animal cells. In the inner part, the mitochondrial matrix has ribosomes, many oxidative enzymes, and a circular DNA molecule that contains the genetic information for some of these enzymes.

Mitosis: A process by which the cell nucleus is divided into two daughter nuclei, each with the same genetic function as the primitive cell. It has four stages: prophase, metaphase, anaphase, and telophase. The division of the cell usually takes place immediately after the division of the nucleus during the mitotic telophase.

Mortality: Death occurrence that is studied in a given population or subpopulation. The word mortality is usually used in an incorrect manner instead of mortality index.

Mutation: Any relatively stable hereditary change of the genetic material that may be a chemical transformation of an individual gene (genetic or punctual mutation) that alters its function or that generates chromosome reordering, gain, or loss that is visible with the microscope. The mutations can occur in germinal cells and can be passed on to the descendants, or they can occur in somatic cells and can be passed on from one cell to another after their division.

Mutagen: Any substance that can generate hereditary changes (mutations) in the genotype of a cell as a consequence of gene or chromosome alterations or loss.

Necrosis: Massive death of tissue areas surrounded by healthy zones. Morphological changes subsequent to cell death, frequently characterized by nuclear changes.

Neoplasia: New and abnormal formation of tumor tissue, or the growth of cell proliferation that is faster than normal and that continues after the initial stimulation that triggered it stopped.

Harmful: Agent that, after contact or absorption, can cause disease or adverse effects in the current or future generations right at the time of the exposure or afterwards.

Odds ratio: Term used in Statistics. It is the quotient obtained after dividing a set of "odds" by another set.

Osteoporosis: Significant reduction of the bone mass with increase of porosity, and there are more tendencies to fractures.

Pesticide: In a strict sense, substance that kills plagues; in an ordinary sense, any substance that is used to take control, avoid, or destroy animal, microbial, or plant plagues.

Population at risk: A group of people who can generate an adverse effect and who are potentially exposed to a determined risk factor. Those people who have already acquired a disease are excluded from the incidence analysis.

Polymorphism: Variation in the sequence of a determined DNA place among individuals from a community. Those polymorphisms that affect the codificant or regulating sequence and that generate important changes in the protein structure or in the expression regulation mechanism may be translated into different phenotypes.

Genetic polymorphism: A situation in which a genetic character appears in more than one way in a community, causing coexistence of more than a morphological type.

Population: A group of individuals of the same species that live in the same geographic area.

Psychosis: Mental disorder characterized by personality disturbance and loss of contact with real facts.

Recycling (waste): It is a process that makes it possible to recover part of the waste for reusable material or for energy.

Risk: Probability of adverse effects or damage taking place for exposure to a toxic agent, due to the inherent properties of the agent and the circumstances or exposure level.

Rhinitis: Inflammation of nasal mucus membranes.

Healthy: Total physical, mental, and social good condition, rather than only the absence of pain or disease. It is a dynamic balanced condition in which an individual or a group is capable of facing any circumstance. Perfect condition.

Safety: The opposite of risk; a certain action that, in defined conditions, damage will not be generated from a dangerous situation. In Toxicology, high probability that the exposure to a substance, in defined conditions of quantity and form that minimize the exposure, will not cause damage.

Significance degree (p): In a comparative study, the significance degree values the authenticity of a hypothesis in contrast with the empirical data. By agreement, it is considered significant (that disagrees with the hypothesis) any diversion with a significance degree of $p < 0.05$, which leads to rejection of the hypothesis.

Syndrome: A set of signs and symptoms that characterizes a determined disease.

Symptom: Subjective evidence of condition or disease perceived by the individual who experiences it.

Nervous system: A set of nerves, nervous centers, tissue, and ganglions. There are sensitive and nervous nerves.

Central nervous system: It is made up of the encephalon (brain and cerebellum) and the spinal cord (inside the spine), connected by the cerebral trunk (medulla oblongata).

Susceptibility: A condition whereby there is a decrease of resistance in an individual before a determined disease or intoxication. This condition is experienced with lower dose exposure than the normally harmful dose for the rest of the population.

Teratogen: An agent that induces structural malformations or defects for the descendents, by means of the mother in prenatal period.

Toxicity: Capacity to cause damage to a living being, in relation to the amount or dose of a substance taken or absorbed, the way it was taken, and its distribution in time (once or repeated times), type and damage seriousness, necessary time to produce it, the nature of the affected organism, and other involving conditions.

Toxic: Any chemical or physical agent that is capable of producing an adverse effect on health. All the physical and chemical agents are potentially toxic, since their action depends on the dose and the individual and environmental circumstances.

Toxicology: Scientific discipline focused on the study of the current or potential danger on the part of the harmful effects of the chemical substances on living beings and ecosystems, the study of the relations of said harmful effects with the exposure, and the study of the mechanisms of action, diagnosis, prevention, and treatment for intoxications.

Transcription: A process by which the genetic information, codified in a linear sequence of nucleotide in a DNA branch, is transferred to an exactly complementary RNA sequence.

Reverse transcription: A process by which the synthesis of a DNA chain from a RNA chain is carried out.

Tumor: Inflammation or abnormal growth of a tissue, regardless of whether it is benign or malignant.

Ulcer: Local loss of substance that often comes with inflammation that affects the surface of an organ or a tissue, caused by the detachment of a necrosis zone and with scarce or non-existent cicatrization capacity.

Air circulation: Fresh air provision to a room or a building.

INDEX

A

B

C

D

E

F

G

H

I

K

L

M

N

O

P

R

S

T

X